GEOMETRY
SUCCESS
IN 20 MINUTES
A DAY

3rd Edition

LEARNINGEXPRESS®

NEW YORK

OTHER TITLES OF INTEREST FROM
LEARNINGEXPRESS

Algebra Success in 20 Minutes a Day
Practical Math Success in 20 Minutes a Day
Statistics Success in 20 Minutes a Day
Trigonometry Success in 20 Minutes a Day

Library of Congress Cataloging-in-Publication Data:
    Geometry success : in 20 minutes a day—3rd ed.
        p. cm.
    ISBN-13: 978-1-57685-745-8 (pbk. : alk. paper)
    ISBN-10: 1-57685-745-X (pbk. : alk. paper) 1. Geometry—programmed instruction.
    1. LearningExpress (Organization)

    QA461.T46 2010
    516—dc22

                                    2009053260

Printed in the United States of America
9 8 7 6 5 4 3 2 1

Third Edition

ISBN-10 1-57685-745-X
ISBN-13 978-1-57685-745-8

For more information or to place an order, contact LearningExpress at:
    LearningExpress
    2 Rector Street
    26th Floor
    New York, NY 10006

Or visit us at:
    www.learnatest.com

CONTENTS ▶

CONTENTS

CONTENTS

INTRODUCTION ▶

This book will help you achieve success in geometry. Reading about math is often slower than reading for amusement. The only assignment more difficult than working math problems is reading about math problems, so numerous figures and illustrations are included to help you understand the material. Although the title of this book suggests studying each lesson for 20 minutes a day, you should work at your own pace through the lessons.

This book is the next best thing to having your own private tutor. It addresses the kinds of questions students have about geometry, and answers those questions in a clear and understandable way. As you work through the lessons in this book, you should feel as if someone is guiding you through each one.

How to Use This Book

Geometry Success in 20 Minutes a Day teaches basic geometry concepts in 20 self-paced lessons. The book also includes a pretest, a posttest, a glossary of mathematical terms, an appendix with postulates and theorems, and an appendix of additional resources for further study. Before you begin Lesson 1, take the pretest, which will assess your current knowledge of geometry. You'll find the answers in the answer key at the end of the book. Taking the pretest will help you determine your strengths and weaknesses in geometry. After taking the pretest, move on to Lesson 1.

Lessons 1–19 offer detailed explanations of basic geometry topics, and Lesson 20 introduces basic trigonometry. Each lesson includes example problems with step-by-step solutions. After you study the examples, you're given a chance to practice similar problems. The answers to the practice problems are in the answer key located at the back of the book. At the end of each lesson is an exercise called Skill Building until Next Time. This exercise applies the lesson's topic to an activity you may encounter in your daily life since geometry is a tool that is used to solve many real-life problems.

After you have completed all 20 lessons, take the posttest, which has the same format as the pretest but different questions. Compare your scores to see how much you've improved or to identify areas in which you need more practice.

If you feel that you need more help with geometry after you complete this book, see Appendix B for additional resources to help you continue improving your geometry skills.

Make a Commitment

Success in geometry requires effort. Make a commitment to improve your geometry skills. Work for understanding. *Why* you do a math operation is as important as *how* you do it. If you truly want to be successful, make a commitment to spend the time you need to do a good job. You can do it! When you achieve success in geometry, you will have laid a solid foundation for future challenges and successes.

So sharpen your pencil and get ready to begin the pretest!

PRETEST

Before you begin the first lesson, you may want to find out how much you already know and how much you need to learn. If that's the case, take the pretest in this chapter, which includes 50 multiple-choice questions covering the topics in this book. While 50 questions can't cover every geometry skill taught in this book, your performance on the pretest will give you a good indication of your strengths and weaknesses.

If you score high on the pretest, you have a good foundation and should be able to work your way through the book quickly. If you score low on the pretest, don't despair. This book will take you through the geometry concepts, step-by-step. If you get a low score, you may need more than 20 minutes a day to work through a lesson. However, this is a self-paced program, so you can spend as much time on a lesson as you need. You decide when you fully comprehend the lesson and are ready to go on to the next one.

Take as much time as you need to complete the pretest. When you are finished, check your answers in the answer key at the end of the book. Each answer also tells you which lesson of this book teaches you about the geometry skills needed for that question.

1.	ⓐ	ⓑ	ⓒ	ⓓ	18.	ⓐ	ⓑ	ⓒ	ⓓ	35.	ⓐ	ⓑ	ⓒ	ⓓ
2.	ⓐ	ⓑ	ⓒ	ⓓ	19.	ⓐ	ⓑ	ⓒ	ⓓ	36.	ⓐ	ⓑ	ⓒ	ⓓ
3.	ⓐ	ⓑ	ⓒ	ⓓ	20.	ⓐ	ⓑ	ⓒ	ⓓ	37.	ⓐ	ⓑ	ⓒ	ⓓ
4.	ⓐ	ⓑ	ⓒ	ⓓ	21.	ⓐ	ⓑ	ⓒ	ⓓ	38.	ⓐ	ⓑ	ⓒ	ⓓ
5.	ⓐ	ⓑ	ⓒ	ⓓ	22.	ⓐ	ⓑ	ⓒ	ⓓ	39.	ⓐ	ⓑ	ⓒ	ⓓ
6.	ⓐ	ⓑ	ⓒ	ⓓ	23.	ⓐ	ⓑ	ⓒ	ⓓ	40.	ⓐ	ⓑ	ⓒ	ⓓ
7.	ⓐ	ⓑ	ⓒ	ⓓ	24.	ⓐ	ⓑ	ⓒ	ⓓ	41.	ⓐ	ⓑ	ⓒ	ⓓ
8.	ⓐ	ⓑ	ⓒ	ⓓ	25.	ⓐ	ⓑ	ⓒ	ⓓ	42.	ⓐ	ⓑ	ⓒ	ⓓ
9.	ⓐ	ⓑ	ⓒ	ⓓ	26.	ⓐ	ⓑ	ⓒ	ⓓ	43.	ⓐ	ⓑ	ⓒ	ⓓ
10.	ⓐ	ⓑ	ⓒ	ⓓ	27.	ⓐ	ⓑ	ⓒ	ⓓ	44.	ⓐ	ⓑ	ⓒ	ⓓ
11.	ⓐ	ⓑ	ⓒ	ⓓ	28.	ⓐ	ⓑ	ⓒ	ⓓ	45.	ⓐ	ⓑ	ⓒ	ⓓ
12.	ⓐ	ⓑ	ⓒ	ⓓ	29.	ⓐ	ⓑ	ⓒ	ⓓ	46.	ⓐ	ⓑ	ⓒ	ⓓ
13.	ⓐ	ⓑ	ⓒ	ⓓ	30.	ⓐ	ⓑ	ⓒ	ⓓ	47.	ⓐ	ⓑ	ⓒ	ⓓ
14.	ⓐ	ⓑ	ⓒ	ⓓ	31.	ⓐ	ⓑ	ⓒ	ⓓ	48.	ⓐ	ⓑ	ⓒ	ⓓ
15.	ⓐ	ⓑ	ⓒ	ⓓ	32.	ⓐ	ⓑ	ⓒ	ⓓ	49.	ⓐ	ⓑ	ⓒ	ⓓ
16.	ⓐ	ⓑ	ⓒ	ⓓ	33.	ⓐ	ⓑ	ⓒ	ⓓ	50.	ⓐ	ⓑ	ⓒ	ⓓ
17.	ⓐ	ⓑ	ⓒ	ⓓ	34.	ⓐ	ⓑ	ⓒ	ⓓ					

Pretest

1. Which of the following sets contains noncollinear points?

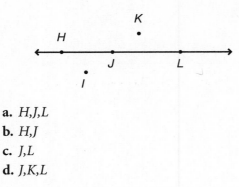

a. H,J,L
b. H,J
c. J,L
d. J,K,L

2. Which is a correct name for this line?

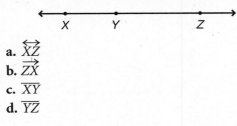

a. $\overleftrightarrow{XZ}$
b. $\overrightarrow{ZX}$
c. $\overline{XY}$
d. $\overline{YZ}$

3. Which is not a property of a plane?
a. is a flat surface
b. has no thickness
c. has boundaries
d. has two dimensions

4. Which is a correct name for this angle?

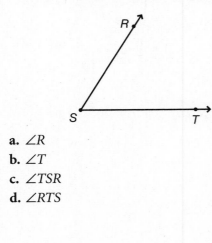

a. ∠R
b. ∠T
c. ∠TSR
d. ∠RTS

5. Which is NOT a correct name for the angle?

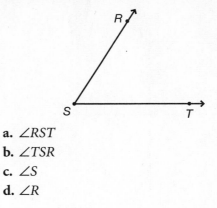

a. ∠RST
b. ∠TSR
c. ∠S
d. ∠R

6. Which line is a transversal?

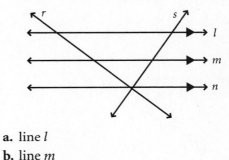

a. line *l*
b. line *m*
c. line *n*
d. line *r*

7. Which pairs of angles are congruent?

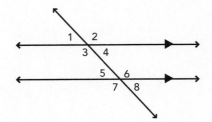

a. ∠1 and ∠2
b. ∠1 and ∠5
c. ∠1 and ∠3
d. ∠1 and ∠6

8. What is the best way to describe the pair of lines *l* and *m*?

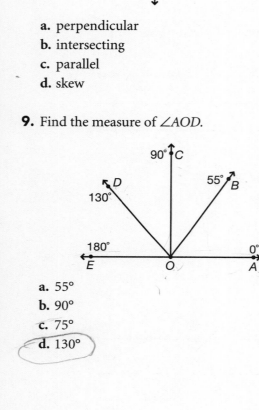

   a. perpendicular
   b. intersecting
   c. parallel
   d. skew

9. Find the measure of ∠AOD.

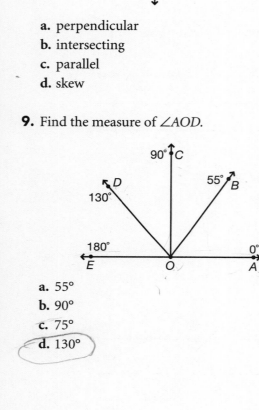

   a. 55°
   b. 90°
   c. 75°
   d. 130°

10. Find the measure of ∠BOD.

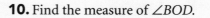
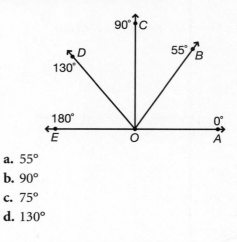

   a. 55°
   b. 90°
   c. 75°
   d. 130°

11. What is the measure of ∠RUS?

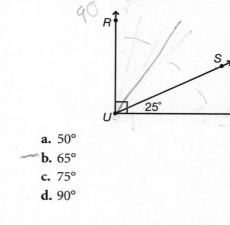

   a. 50°
   b. 65°
   c. 75°
   d. 90°

12. What is the measure of ∠JTK?

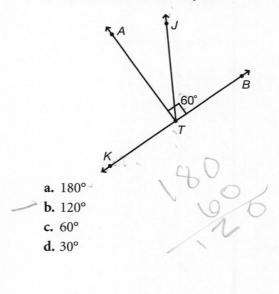

   a. 180°
   b. 120°
   c. 60°
   d. 30°

13. What angle would be supplementary to *ONP*?

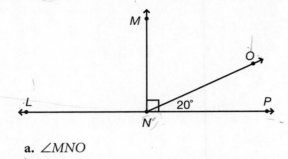

   a. ∠*MNO*
   b. ∠*MNP*
   c. ∠*MNL*
   d. ∠*LNO*

14. Classify the triangle by its sides.

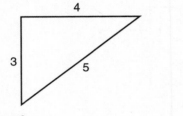

   a. scalene
   b. isosceles
   c. equilateral
   d. none of the above

15. Classify △*ABC* with the following measurements, *AB* = 3, *BC* = 7, and *AC* = 7.
   a. scalene
   b. isosceles
   c. equilateral
   d. none of the above

16. △*SAW* is an isosceles triangle. What is the best name for ∠*W*?

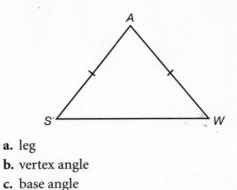

   a. leg
   b. vertex angle
   c. base angle
   d. base

17. ∠*FEH* is approximately equal ∠_____?

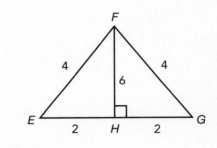

   a. ∠*FGE*
   b. ∠*FGH*
   c. ∠*HEF*
   d. ∠*HGF*

18. Which postulate could you use to prove △*FGH* ≅ △*PQR*?

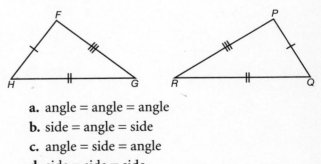

   a. angle = angle = angle
   b. side = angle = side
   c. angle = side = angle
   d. side = side = side

19. The hypotenuse of △*HLM* is

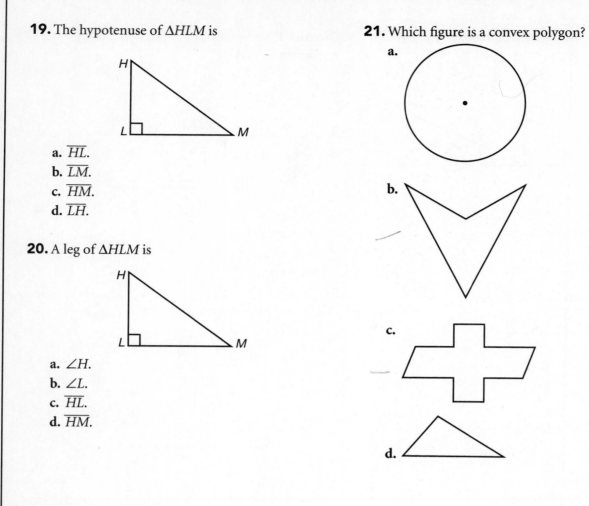

    a. $\overline{HL}$.
    b. $\overline{LM}$.
    c. $\overline{HM}$.
    d. $\overline{LH}$.

20. A leg of △*HLM* is

    a. ∠*H*.
    b. ∠*L*.
    c. $\overline{HL}$.
    d. $\overline{HM}$.

21. Which figure is a convex polygon?

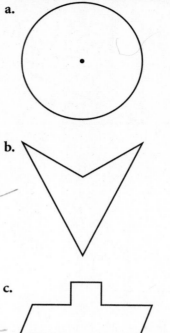

    a.

    b.

    c.

    d.

22. What is the sum of the measures of the interior angles of a convex quadrilateral?
a. 45°
b. 90°
c. 180°
d. 360°

23. What is the name for a polygon with six sides?
a. pentagon
b. hexagon
c. octagon
d. decagon

24. Which of the following is not necessarily a parallelogram?
a. quadrilateral
b. rectangle
c. rhombus
d. square

25. Which of the following is NOT a quadrilateral?
a. trapezoid
b. parallelogram
c. decagon
d. square

26. Express the ratio $\frac{JL}{JM}$ in simplest form.

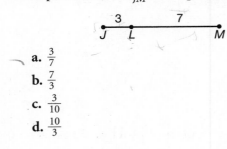

a. $\frac{3}{7}$
b. $\frac{7}{3}$
c. $\frac{3}{10}$
d. $\frac{10}{3}$

27. Solve for y: $\frac{5}{y} = \frac{20}{4}$
a. 1
b. 4
c. 5
d. 20

28. Which of the following is a true proportion?
a. 3:7 = 5:8
b. 1:2 = 4:9
c. 6:3 = 10:6
d. 2:5 = 4:10

29. Find the perimeter of the polygon.

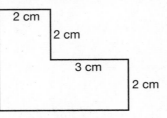

a. 9 cm
b. 18 cm
c. 36 cm
d. 72 cm

30. Find the perimeter of a square that measures 16 inches on one side.
a. 16 in.
b. 32 in.
c. 64 in.
d. not enough information

31. Find the area of a rectangle with base 7 inches and height 11 inches.
a. 77 in.^{2}
b. 36 in.^{2}
c. 18 in.^{2}
d. 154 in.^{2}

32. Find the area of a parallelogram with base 5 cm and height 20 cm.
  a. 100 cm²
  b. 50 cm²
  c. 25 cm²
  d. 15 cm²

33. Find the area of the trapezoid in the figure.

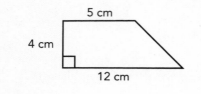

  a. 18 cm²
  b. 24 cm²
  c. 34 cm²
  d. 60 cm²

34. Which segment does not equal 3 m?

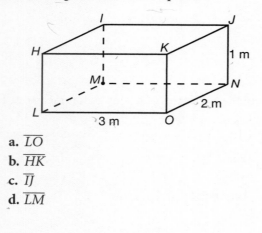

  a. $\overline{LO}$
  b. $\overline{HK}$
  c. $\overline{IJ}$
  d. $\overline{LM}$

35. Use the formula $SA = 2(lw + wh + lh)$ to find the surface area of the prism.

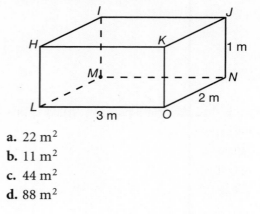

  a. 22 m²
  b. 11 m²
  c. 44 m²
  d. 88 m²

36. Find the volume of a prism with length 12 inches, width 5 inches, and height 8 inches.
  a. 60 in.³
  b. 40 in.³
  c. 480 in.³
  d. 96 in.³

37. Find the volume of the triangular prism.

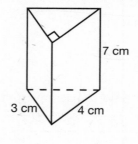

  a. 14 cm³
  b. 21 cm³
  c. 42 cm³
  d. 84 cm³

38. Find the volume of a pyramid whose base has an area of 10 in.² and whose height is 6 in.
  a. 10 in.³
  b. 20 in.³
  c. 30 in.³
  d. 60 in.³

39. Find the circumference of a circle with a diameter of 21 inches. Use 3.14 for π.
   a. 32.47 in.
   b. 129.88 in.
   c. 756.74 in.
   d. 65.94 in.

40. Find the area of a circle with a diameter of 20 cm. Use 3.14 for π.
   a. 628 cm^2
   b. 62.8 cm^2
   c. 314 cm^2
   d. 31.4 cm^2

41. In which quadrant would you graph the point $(5,-6)$?
   a. I
   b. II
   c. III
   d. IV

42. The coordinates for point A are

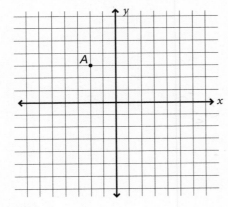

   a. $(3,-2)$.
   b. $(-2,3)$.
   c. $(-3,2)$.
   d. $(2,-3)$.

43. Find the slope of a line that passes through the points $(1,1)$ and $(-3,5)$.
   a. 1
   b. -1
   c. $\frac{1}{4}$
   d. $-\frac{1}{4}$

44. A line that points up to the right has a slope that is
   a. positive.
   b. negative.
   c. zero.
   d. undefined.

45. A vertical line has a slope that is
   a. positive.
   b. negative.
   c. zero.
   d. undefined.

46. Which of the following is a linear equation?
   a. $\frac{10}{x} = y$
   b. $3x + 2y^2 = 10$
   c. $3x^2 + 2y = 10$
   d. $3x + 2y = 10$

47. Which of the following is not a linear equation?
   a. $x = 4$
   b. $y = -4$
   c. $\frac{1}{2}x + \frac{1}{3}y = 7$
   d. $\frac{2}{x} = y$

48. Which ordered pair satisfies the equation
$3x + 4y = 12$?
   a. (0,0)
   b. (2,2)
   c. (1,4)
   d. (0,3)

49. The ratio of the opposite leg to an adjacent leg is the trigonometric ratio
   a. sine.
   b. cosine.
   c. tangent.
   d. hypotenuse.

50. The ratio of the adjacent leg to the hypotenuse is the trigonometric ratio
   a. sine.
   b. cosine.
   c. tangent.
   d. hypotenuse.

THE BASIC BUILDING BLOCKS OF GEOMETRY

Lesson Summary

This lesson explains the basic building blocks of geometry: points, lines, rays, line segments, and planes. It also shows you the basic properties you need to understand and apply these terms.

The term *geometry* is derived from the two Greek words *geo* and *metron*. It means "to measure the Earth." The great irony is that the most basic building block in geometry, the *point*, has no measurement at all. But you must accept that a point exists in order to have lines and planes, because lines and planes are made up of an infinite number of points. Let's begin this lesson by looking at each of the basic building blocks of geometry.

Points

A *point* has no size and no dimension; however, it indicates a definite location. Some real-life examples are a pencil point, a corner of a room, and the period at the end of this sentence. A series of points is what makes up lines, line segments, rays, and planes. There are countless points on any one line. A point is named with an italicized uppercase letter placed next to it:

$$\cdot A$$

If you want to discuss this point with someone, you would call it "point *A*."

Lines

A *line* is straight, but it has no thickness. It is an infinite set of points that extends in both directions. Imagine a straight highway with no end and no beginning; this is an example of a line. A line is named by any one italicized lowercase letter or by naming any two points on the line. If the line is named by using two points on the line, a small symbol of a line (with two arrows) is written above the two letters. For example, this line could be referred to as line $\overleftrightarrow{AB}$ or line $\overleftrightarrow{BA}$:

$$\longleftrightarrow \overset{A}{\bullet} \qquad \overset{B}{\bullet} \longrightarrow$$

Practice

The answer key for practice exercises begins on p. 194.

1. Are there more points on $\overleftrightarrow{AB}$ than point *A* and point *B*?

2. How are points distinguished from one another?

3. Why would lines, segments, rays, or planes not exist if points do not exist?

4. Write six different names for this line. $\longleftrightarrow \overset{X}{\bullet} \quad \overset{Y}{\bullet} \ \overset{Z}{\bullet} \longrightarrow$

5. How many points are on a line?

6. Why do you think the notation for a line has two arrowheads?

Rays

A *ray* is a part of a line with one endpoint that continues indefinitely in the direction opposite the endpoint. It has an infinite number of points on it. Flashlights and laser beams are good ways to imagine rays. When you refer to a ray, you always name the endpoint first. The ray shown here is ray $\overrightarrow{AB}$, not ray $\overrightarrow{BA}$.

Line Segments

A *line segment* is part of a line with two endpoints. It has an infinite number of points on it. A ruler and a base-board are examples of ways you might picture a line segment. Like lines and rays, line segments are also named with two italicized uppercase letters, but the symbol above the letters has no arrows. This segment could be referred to as $\overline{AB}$ or $\overline{BA}$:

Practice

7. Name two different rays with endpoint *S*.

8. Why is it important to name the endpoint of a ray first?

9. Why are ray $\overrightarrow{AB}$ and ray $\overrightarrow{BA}$ not the same?

10. Name six different line segments shown.

11. Why are arrowheads not included in line segment notation?

12. How many points are on a line segment?

Planes

A *plane* is a flat surface that has no thickness or boundaries. Imagine a floor that extends in all directions as far as you can see. Here is what a plane looks like:

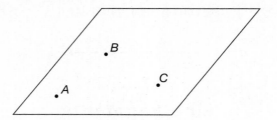

When you talk to someone about this plane, you could call it "plane *ABC*." However, the more common practice is to name a plane with a single italicized capital letter (no dot after it) placed in the upper-right corner of the figure as shown here:

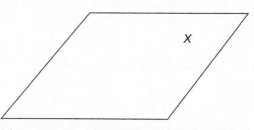

If you want to discuss this plane with someone, you would call it "plane *X*."

THE BASIC BUILDING BLOCKS OF GEOMETRY

FIGURE	NAME	SYMBOL	READ AS	PROPERTIES	EXAMPLES
• *A*	Point	• *A*	point *A*	• has no size • has no dimension • indicates a definite location • named with an italicized uppercase letter	• pencil point • corner of a room
A B ←→ *l* ←→	Line	$\overleftrightarrow{AB}$ or $\overleftrightarrow{BA}$	line *AB* or *BA* or line *l*	• is straight • has no thickness • an infinite set of points that extends in opposite directions • one dimension	• continuous highway without boundaries • hallway without ends
A B —→	Ray	$\overrightarrow{AB}$	ray *AB* (endpoint is always first)	• is part of a line • has only one endpoint • an infinite set of points that extends in one direction • one dimension	• flashlight • laser beam
A B ——	Line segment	$\overline{AB}$ or $\overline{BA}$	segment *AB* or *BA*	• is part of a line • has two endpoints • an infinite set of points • one dimension	• edge of a ruler • baseboard
(plane figure with points *B*, *A*, *C* and plane *X*)	Plane	None	plane *ABC* or plane *X*	• is a flat surface • has no thickness • an infinite set of points that extends in all directions • two dimensions	• floor without boundaries • surface of a football field without boundaries

Practice

13. A line is different from a ray because _____.

14. A property of a point is _____.

15. A ray is different from a segment because _____.

16. A property of a line segment is _____.

17. A plane is different from a line because _____.

Working with the Basic Building Blocks of Geometry

Points, lines, rays, line segments, and planes are very important building blocks in geometry. Without them, you cannot work many complex geometry problems. These five items are closely related to one another. You will use them in all the lessons that refer to *plane figures*—figures that are flat with one or two dimensions. Later in the book, you will study three-dimensional figures—figures that occur in space. *Space* is the set of all possible points and is three-dimensional. For example, a circle and a square are two-dimensional and can occur in a plane. Therefore, they are called *plane figures*. A sphere (ball) and a cube are examples of three-dimensional figures that occur in space, not a plane.

One way you can see how points and lines are related is to notice whether points lie in a straight line. *Collinear points* are points on the same line. *Noncollinear* points are points not on the same line. Even though you may not have heard these two terms before, you probably understand the following two figures based on the sound of the names *collinear* and *noncollinear*.

Collinear points

Noncollinear points

A way to see how points and planes are related is to notice whether points lie in the same plane. For instance, see these two figures:

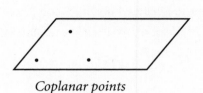

Coplanar points

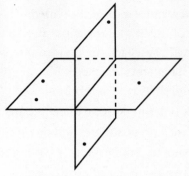

Noncoplanar points

Again, you may have guessed which name is correct by looking at the figures and seeing that in the figure labeled *Coplanar points,* all the points are on the same plane. In the figure labeled *Noncoplanar points,* the points are on different planes.

Practice

18. What are three examples of figures that occur in space?

19. Can three points be noncollinear? Why or why not?

20. Can coplanar points be noncollinear? Why or why not?

21. Can collinear points be coplanar? Why or why not?

Postulates and Theorems

You need a few more tools before moving on to other lessons in this book. Understanding the terms of geometry is only part of the battle. You must also be able to understand and apply certain facts about geometry and geometric figures. These facts are divided into two categories: postulates and theorems. *Postulates* (sometimes called *axioms*) are statements accepted without proof. *Theorems* are statements that can be proved. You will be using both postulates and theorems in this book, but you will not be proving the theorems.

Geometry is the application of definitions, postulates, and theorems. Euclid is known for compiling all the geometry of his time into postulates and theorems. His masterwork, *The Elements*, written about 300 B.C., is the basis for many geometry books today. We often refer to this as Euclidean geometry.

Here are two examples of postulates:

> *Postulate:* Two points determine exactly one line.
>
> *Postulate:* Three noncollinear points determine exactly one plane.

Practice

State whether each set of points is collinear.

22. *A, B, C*

23. *A, E, F*

24. *B, D, F*

25. *A, E*

Determine whether each set of points is coplanar.

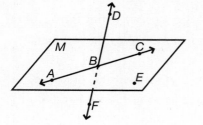

26. *A, B, C, E*

27. *D, B, C, E*

28. *B, C, E, F*

29. *A, B, E*

Determine whether the following statements are true or false.

30. $\overleftrightarrow{XY}$ and $\overleftrightarrow{YX}$ are the same line.

31. $\overrightarrow{XY}$ and $\overrightarrow{YX}$ are the same ray.

32. $\overline{XY}$ and $\overline{YX}$ are the same segment.

33. Any four points *W, X, Y,* and *Z* must lie in exactly one plane.

Draw and label a figure for the following two questions to fit each description, if possible. Otherwise, state "not possible."

34. four collinear points

35. three noncoplanar points

36. Are three points that are collinear *sometimes, always,* or *never* coplanar?

Skill Building until Next Time

If you place a four-legged table on a surface and it wobbles, then one of the legs is shorter than the other three. You can use three legs to support something that must be kept steady. Why do you think this is true?

Throughout the day, be on the lookout for some three-legged objects that support this principle. Examples include a camera tripod and an artist's easel. The bottoms of the three legs must represent three noncollinear points and determine exactly one plane.

LESSON

2 ▶ TYPES OF ANGLES

Lesson Summary

This lesson will teach you how to classify and name several types of angles. You will also learn about opposite rays.

Many of us use the term *angle* in everyday conversations. For example, we talk about camera angles, angles for pool shots and golf shots, and angles for furniture placement. In geometry, an *angle* is formed by two rays with a common endpoint. The symbol used to indicate an angle is ∠. The two rays are the sides of the angle. The common endpoint is the vertex of the angle. In the following figure, the sides are $\overrightarrow{RD}$ and $\overrightarrow{RY}$, and the vertex is *R*.

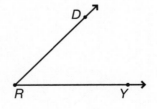

Naming Angles

People call you different names at different times. Sometimes, you are referred to by your full name, and other times, only by your first name or maybe even a nickname. These different names don't change who you are—just the way in which others refer to you. You can be named differently according to the setting you're in. For example, you may be called your given name at work, but your friends might call you by a nickname. Confusion can sometimes arise when these names are different.

Just like you, an angle can be named in different ways. The different ways an angle can be named may be confusing if you do not understand the logic behind the different methods of naming.

If three letters are used to name an angle, then the middle letter always names the vertex. If the angle does not share the vertex point with another angle, then you can name the angle with only the letter of the vertex. If a number is written inside the angle that is not the angle measurement, then you can name the angle by that number. You can name the following angle any one of these names: ∠WET, ∠TEW, ∠E, or ∠1.

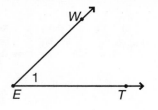

Practice

1. Name the vertex and sides of the angle.

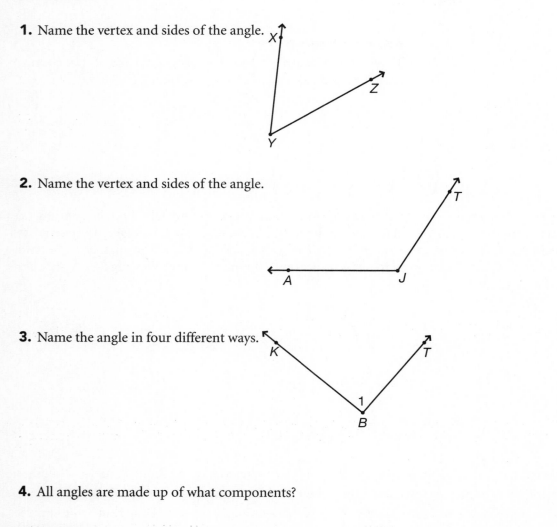

2. Name the vertex and sides of the angle.

3. Name the angle in four different ways.

4. All angles are made up of what components?

Right Angles

Angles that make a square corner are called *right angles* (see p. 26 for more details about what makes an angle a right angle). In drawings, the following symbol is used to indicate a right angle:

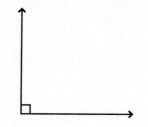

Straight Angles

Opposite rays are two rays with the same endpoint that form a line. They form a *straight angle.* In the following figure, $\overrightarrow{HD}$ and $\overrightarrow{HS}$ are opposite rays.

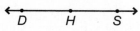

Practice

Use the following figure to answer practice problems 5–8.

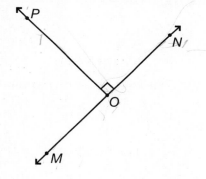

5. Name two right angles.

6. Name a straight angle.

7. Name a pair of opposite rays.

8. Why is $\angle O$ an incorrect name for any of the angles shown?

Use the following figure to answer practice problems 9 and 10.

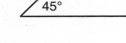

9. Are $\overrightarrow{LM}$ and $\overrightarrow{MN}$ opposite rays? Why or why not?

10. If two rays share the same endpoint, do they have to be opposite rays? Why or why not?

Classifying Angles

Angles are often classified by their measures. The degree is the most commonly used unit for measuring angles. One full turn, or a circle, equals 360°.

Acute Angles

An *acute angle* has a measure between 0° and 90°. Here are two examples of acute angles:

45°

89°

Right Angles

A *right angle* has a 90° measure. The corner of a piece of paper will fit exactly into a right angle. Here are two examples of right angles:

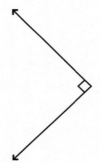

Obtuse Angles

An *obtuse angle* has a measure between 90° and 180°. Here are two examples of obtuse angles:

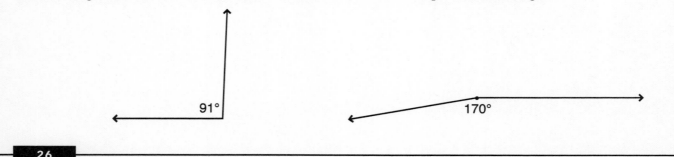

91°

170°

Straight Angles

A straight angle has a 180° measure. ∠ABC is an example of a straight angle:

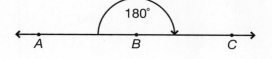

Practice

Use the following figure to answer practice problems 11–15.

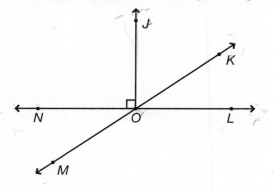

11. Name three acute angles.

12. Name three obtuse angles.

13. Name two straight angles.

14. Determine the largest angle shown, and write it two ways.

15. If ∠MON measures 27°, then ∠JOK measures _____ degrees.

Complete each statement.

16. An angle with measure 90° is called a(n) _____ angle.

17. An angle with measure 180° is called a(n) _____ angle.

18. An angle with a measure between 0° and 90° is called a(n) _____ angle.

19. An angle with a measure between 90° and 180° is called a(n) _____ angle.

20. Two adjacent right angles combine to a(n) _____ angle.

21. A straight angle minus an acute angle would result in a(n) _____ angle.

Questions 22–25 list the measurement of an angle. Classify each angle as acute, right, obtuse, or straight.

22. 10°

24. 98°

23. 180°

25. 90°

You may want to use a corner of a piece of scratch paper for questions 26–29. Classify each angle as acute, right, obtuse, or straight.

26.

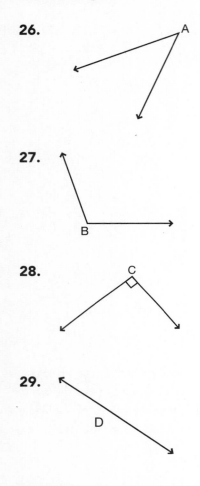

27.

28.

29.

30. Take angles *A*, *B*, *C*, and *D*, from questions 26–29 and list them in order from smallest to largest.

Skill Building until Next Time

In conversational English, the word *acute* can mean sharp and the word *obtuse* can mean dull or not sharp. How can you relate these words with the drawings of acute and obtuse angles?

3 ▶ WORKING WITH LINES

Lesson Summary

This lesson introduces you to perpendicular, transversal, parallel, and skew lines. The angles formed by a pair of parallel lines and a transversal are also explained.

Both intersecting and nonintersecting lines surround you, but you may not pay much attention to them most of the time. In this lesson, you will focus on two different types of intersecting lines: transversals and perpendicular lines. You will also study nonintersecting lines: parallel and skew lines. You will learn about properties of lines that have many applications to this lesson and throughout this book. You'll soon start to look at the lines around you with a different point of view.

Intersecting Lines

On a piece of scratch paper, draw two straight lines that cross. Can you make these straight lines cross at more than one point? No, you can't, because intersecting lines cross at only one point (unless they are the same line). The point where they cross is called a *point of intersection*. They actually share this point, because it is on both lines. Two special types of intersecting lines are called *transversals* and *perpendicular lines*.

Transversals

A *transversal* is a line that intersects two or more other lines, each at a different point. In the following figure, line *t* is a transversal; line *s* is not.

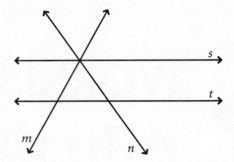

The prefix *trans* means *across*. In the previous figure, you can see that line *t* cuts across the two lines *m* and *n*. Line *m* is a transversal for lines *s* and *t*. Also, line *n* is a transversal across lines *s* and *t*. Line *s* crosses lines *m* and *n* at the same point (their point of intersection); therefore, line *s* is not a transversal. A transversal can cut across parallel as well as intersecting lines, as shown here:

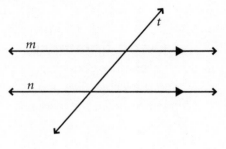

Practice

Use the following figure to answer questions 1–4.

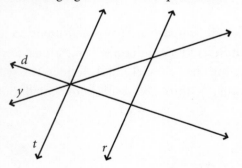

1. Is line *d* a transversal? Why or why not?

2. Is line *y* a transversal? Why or why not?

3. Is line *t* a transversal? Why or why not?

4. Is line *r* a transversal? Why or why not?

Perpendicular Lines

Perpendicular lines are another type of intersecting lines. Everyday examples of perpendicular lines include the horizontal and vertical lines of a plaid fabric and the lines formed by panes in a window. Perpendicular lines meet to form right angles. Right angles always measure 90°. In the following figure, lines x and y are perpendicular:

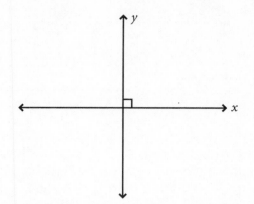

The symbol "⊥" means perpendicular. You could write $x \perp y$ to show these lines are perpendicular. Also, the symbol that makes a square in the corner where lines x and y meet indicates a right angle. In geometry, you shouldn't assume anything without being told. Never assume a pair of lines are perpendicular without one of these symbols. A transversal *can* be perpendicular to a pair of lines, but it does not *have* to be. In the following figure, line t is perpendicular to both line l and line m.

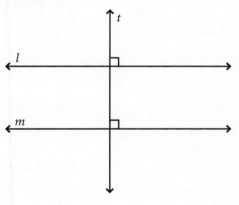

Practice

State whether the following statements are true or false.

_____ **5.** Perpendicular lines always form multiple right angles.

_____ **6.** The symbol "⊥" means parallel.

_____ **7.** Transversals must always be parallel.

_____ **8.** Perpendicular lines can be formed by intersecting or nonintersecting lines.

Nonintersecting Lines

If lines do not intersect, then they are either parallel or skew.

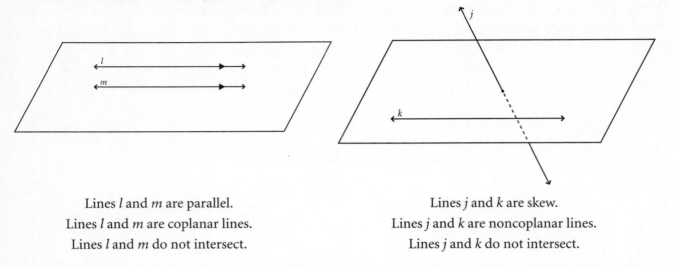

Lines *l* and *m* are parallel.

Lines *l* and *m* are coplanar lines.

Lines *l* and *m* do not intersect.

Lines *j* and *k* are skew.

Lines *j* and *k* are noncoplanar lines.

Lines *j* and *k* do not intersect.

The symbol "∥" means parallel. So you can abbreviate the sentence, "Lines *l* and *m* are parallel," by writing "*l* ∥ *m*." Do not assume a pair of lines are parallel unless it is indicated. Arrowheads on the lines in a figure indicate that the lines are parallel. Sometimes, double arrowheads are necessary to differentiate two sets of parallel lines, as shown in the following figure:

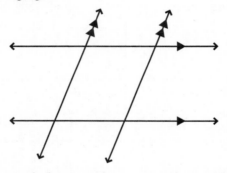

Everyday examples of parallel lines include rows of crops on a farm and railroad tracks. An example of skew lines is the vapor trails of a northbound jet and a westbound jet flying at different altitudes. One jet would pass over the other, but their paths would not cross.

Practice

Complete the sentences with the correct word: *always*, *sometimes*, or *never*.

9. Parallel lines are _____ coplanar.

10. Parallel lines _____ intersect.

11. Parallel lines are _____ cut by a transversal.

12. Lines that are skewed _____ form right angles.

13. Skew lines are _____ coplanar.

14. Skew lines _____ intersect.

Angles Formed by Parallel Lines and a Transversal

If a pair of parallel lines are cut by a transversal, then eight angles are formed. In the following figure, line *l* is parallel to line *m* and line *t* is a transversal forming angles 1–8. Angles 3, 4, 5, and 6 are inside the parallel lines and are called *interior* angles. Angles 1, 2, 7, and 8 are outside the parallel lines and are called *exterior* angles.

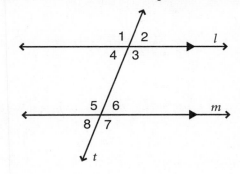

Alternate Interior Angles

Alternate interior angles are interior angles on opposite sides of the transversal. In the previous figure, angles 3 and 5 and angles 4 and 6 are examples of alternate interior angles. To spot alternate interior angles, look for a Z-shaped figure, as shown in the following figures:

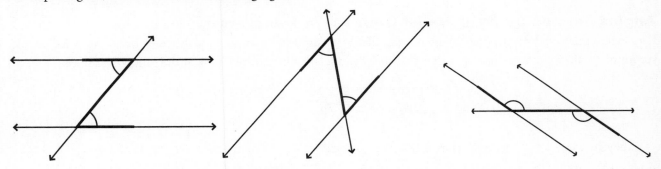

Same-Side Interior Angles

Same-side interior angles are interior angles on the same side of the transversal. To spot same-side interior angles, look for a *U-shaped* figure in various positions, as shown in the following examples:

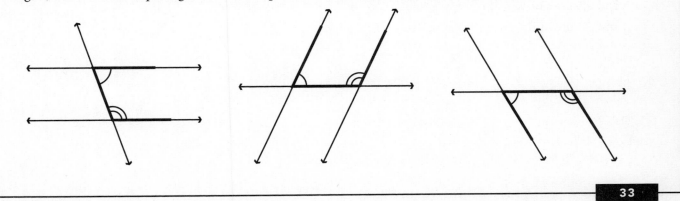

Corresponding Angles

Corresponding angles are so named because they appear to be in corresponding positions in relation to the two parallel lines. Examples of corresponding angles in the following figure are angles 1 and 5, 4 and 8, 2 and 6, and 3 and 7.

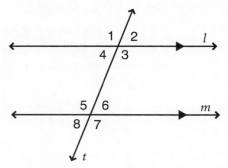

To spot corresponding angles, look for an *F-shaped* figure in various positions, as shown in these examples:

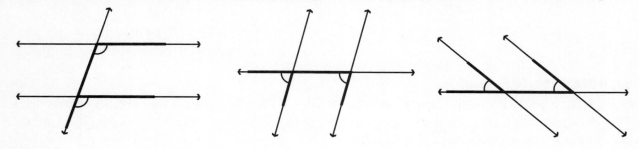

Angles Formed by Nonparallel Lines and a Transversal

Even when lines cut by a transversal are not parallel, you still use the same terms to describe angles, such as corresponding, alternate interior, and same-side interior angles. For example, look at the following figure:

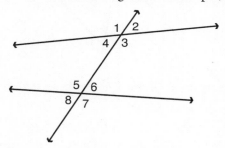

∠1 and ∠5 are corresponding angles.

∠3 and ∠5 are alternate interior angles.

∠4 and ∠5 are same-side interior angles.

Postulates and Theorems

As mentioned in Lesson 1, *postulates* are statements of fact that we accept without proof. *Theorems* are statements of fact that can be proved. You will apply both types of facts to problems in this book. Some geometry books teach formal proofs of theorems. Although you will not go through that process in this book, you will still use both postulates and theorems for applications. There are some important facts you need to know about the special pairs of angles formed by two parallel lines and a transversal. If you noticed that some of those pairs of angles appear to have equal measure, then you are on the right track. The term used for two angles with equal measure is *congruent*.

Another important fact you should know is that a pair of angles whose sum is 180° is called *supplementary*. Here are the theorems and the postulate that you will apply to a figure in the next few practice exercises.

> *Postulate:* If two parallel lines are cut by a transversal, then corresponding angles are congruent.
> *Theorem:* If two parallel lines are cut by a transversal, then alternate interior angles are congruent.
> *Theorem:* If two parallel lines are cut by a transversal, then same-side interior angles are supplementary.

Practice

In the following figure, $m \parallel n$ and $s \parallel t$. For questions 15–18, (a) state the special name for each pair of angles (alternate interior angles, corresponding angles, or same-side interior angles), then (b) tell if the angles are congruent or supplementary.

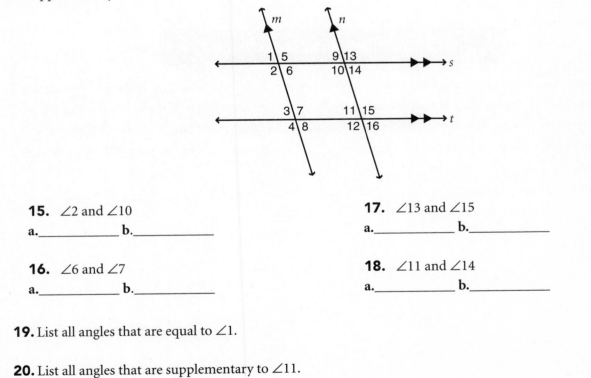

15. ∠2 and ∠10
a._____ b._____

16. ∠6 and ∠7
a._____ b._____

17. ∠13 and ∠15
a._____ b._____

18. ∠11 and ∠14
a._____ b._____

19. List all angles that are equal to ∠1.

20. List all angles that are supplementary to ∠11.

For questions 21–24, use the measure of the given angle to find the missing angle.

21. $m\angle 2 = 100°$, so $m\angle 7 =$ _____

22. $m\angle 8 = 71°$, so $m\angle 12 =$ _____

23. $m\angle 5 = 110°$, so $m\angle 7 =$ _____

24. $m\angle 2 = 125°$, so $m\angle 11 =$ _____

Complete the statement for practice problems 25–30.

25. Alternate interior angles are similar to corresponding angles because _____.

26. Alternate interior angles differ from corresponding angles because _____.

27. Same-side interior angles are similar to alternate interior angles because _____.

28. Same-side interior angles differ from alternate interior angles because _____.

29. Same-side interior angles are similar to corresponding angles because _____.

30. Same-side interior angles differ from corresponding angles because _____.

Skill Building until Next Time

Look around your neighborhood for a building under construction. Are the floor beams the same distance apart at the front of the building as they are at the back? Why is this important? If they were not the same distance apart, how would this effect the floor? Of course, the floor would sag. Building codes specify how far apart floor beams can be based on safety codes. You can see the importance of parallel lines in construction.

▶ MEASURING ANGLES

Lesson Summary

This lesson focuses on using the protractor to measure and draw angles. You will also add and subtract angle measures.

An instrument called a *protractor* can be used to find the measure of degrees of an angle. Most protractors have two scales, one reading from left to right, and the other from right to left. Estimating whether an angle is acute, right, obtuse, or straight before measuring will help you choose the correct scale. Here is an example of a protractor:

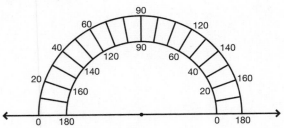

Carefully line up your protractor by placing the center point of the protractor scale on the vertex of the angle, which is the place where both sides of the angle meet. If the sides of the angle do not reach the scale, extend them. Choose the scale that has zero at one side of the angle. Read the measure of the angle. Check to see if your measurement and estimate agree. When measuring an angle, it is not necessary to have one of the rays passing through zero on the protractor scale. The angle could be measured by subtracting the smaller measurement from the larger one. Putting the ray on zero simply makes the counting easier.

Practice

Using the following figure, find the measure of each angle.

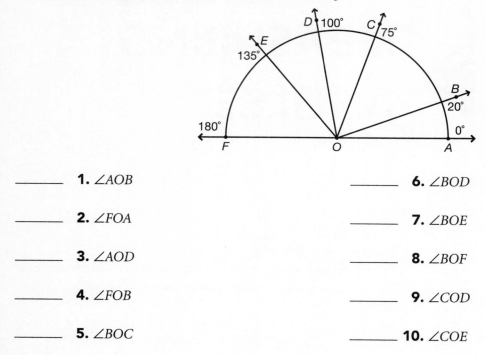

_____ **1.** ∠AOB

_____ **2.** ∠FOA

_____ **3.** ∠AOD

_____ **4.** ∠FOB

_____ **5.** ∠BOC

_____ **6.** ∠BOD

_____ **7.** ∠BOE

_____ **8.** ∠BOF

_____ **9.** ∠COD

_____ **10.** ∠COE

Drawing Angles

You can use a protractor to draw an angle of a given size. First, draw a ray and place the center point of the protractor on the endpoint of the ray. Align the ray with the base line of the protractor. Locate the degree of the angle you wish to draw. Make a dot at that point and connect it to the endpoint of the ray.

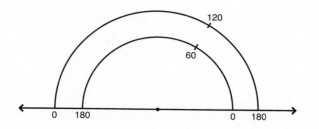

The resulting angle will have the correct degree of measurement:

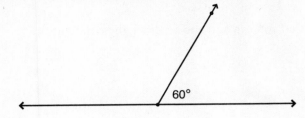

Practice
Use a protractor to draw angles with the given measures.

11. 45°

14. 125°

12. 75°

15. 32°

13. 100°

Adding and Subtracting Angle Measures

The following figures suggest that you can add and subtract angle measures:

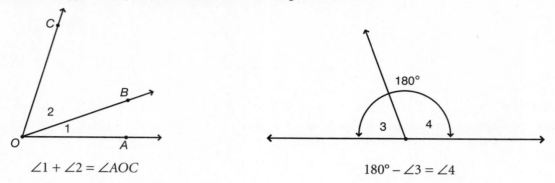

$$\angle 1 + \angle 2 = \angle AOC \qquad\qquad 180° - \angle 3 = \angle 4$$

Adjacent angles are two angles in the same plane that have the same vertex and a common side but do not have any interior points in common.

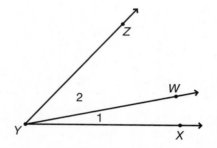

In the previous figure, $\angle 1$ and $\angle 2$ are adjacent angles. Angles 1 and *XYZ* are not adjacent.

Use a protractor to measure $\angle 1$, $\angle 2$, and $\angle XYZ$. What relationship do you notice among the measures of the three angles? You should find that the measure of $\angle 1$ plus the measure of $\angle 2$ equals the measure of $\angle XYZ$. The letter *m* is used before the angle symbol to represent the word *measure*. For example, $m\angle 1 + m\angle 2 = m\angle XYZ$. If you draw another pair of adjacent angles and measure the angles, will the relationship be the same? Try it and see.

Angle Addition Postulate

(1) If point B is in the interior of ∠AOC, then

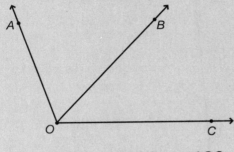

$$m\angle AOB + m\angle BOC = m\angle AOC.$$

(2) If ∠AOC is a straight angle, then

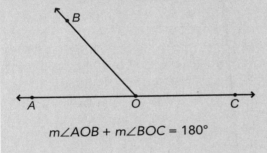

$$m\angle AOB + m\angle BOC = 180°$$

Examples: Find the measure of each angle.

(a) $m\angle 1 = 40°$
     $m\angle 2 = 75°$
     $m\angle RSW =$ _____

(b) $m\angle DEG = 35°$
     $m\angle GEF = 145°$
     $m\angle DEF =$ _____

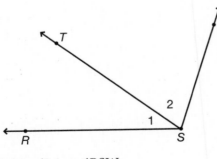

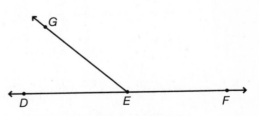

$m\angle 1 + m\angle 2 = m\angle RSW$
$m\angle RSW = 115°$

$m\angle DEG + m\angle GEF = m\angle DEF$
$m\angle DEF = 180°$

Practice

Find the measure of each angle.

16. $m\angle KJB = 60°$

$m\angle BJT = 40°$

$m\angle KJT = $ _____

17. $m\angle DOS = 120°$

$m\angle DOH = 70°$

$m\angle HOS = $ _____

18. $m\angle RPM = 42°$

$m\angle MPS = $ _____

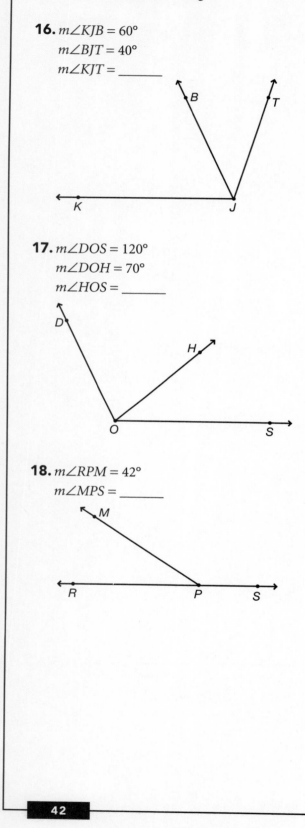

19. $m\angle ACR = 25°$

  $m\angle RCD =$ _____

20. $m\angle GEO = 100°$

  $m\angle GEM = 60°$

  $m\angle MEO =$ _____

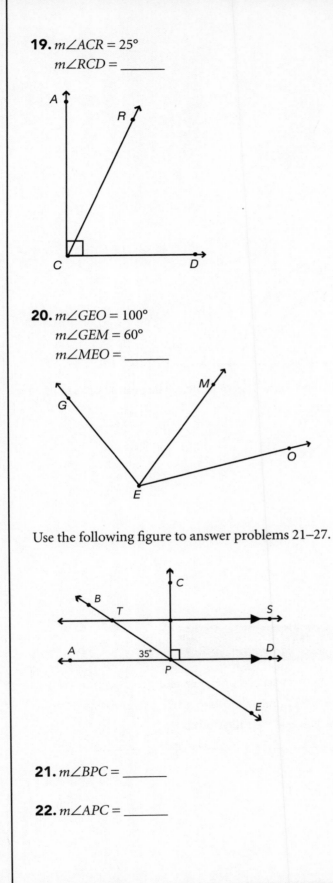

Use the following figure to answer problems 21–27.

21. $m\angle BPC =$ _____

22. $m\angle APC =$ _____

23. $m\angle EPB =$ _____

24. $m\angle CPE =$ _____

25. $m\angle APE =$ _____

26. $m\angle ETS =$ _____

27. $m\angle STB =$ _____

Use the following figure for practice problems 28–31.

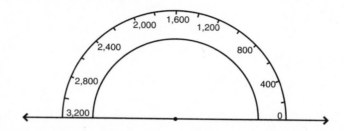

The U.S. Army uses a unit of angle measure called a *mil*. A mil is defined as $\frac{1}{6,400}$ of a circle. The protractor is marked in mils. Find the measure in mils of each of the following angles.

_____ **28.** 90° angle

_____ **29.** 135° angle

_____ **30.** 45° angle

_____ **31.** 180° angle

Skill Building until Next Time

Look around your environment for angles and estimate their measurements. For example, estimate the size of an angle created by leaving your door open partway. You may also want additional practice using your protractor. If so, practice drawing several angles of different measurements.

 PAIRS OF ANGLES

Lesson Summary
In this lesson, you will learn how to use the relationships among three special pairs of angles: complementary angles, supplementary angles, and vertical angles.

hen two angles come together to form a 90° angle, the angle can be very useful for solving a variety of problems. Angles measuring 90° are essential in construction and sewing, just to name two areas.

Pairs of angles whose measures add up to 90° are called *complementary* angles. If you know the measurement of one angle, then you can find the measurement of the other.

When two angles come together to form a 180° angle, or a straight line, the applications are endless. These pairs of angles are called *supplementary* angles. You can find the measurement of one angle if the measurement of the other is given.

When two lines intersect, two pairs of *vertical* angles are always formed. You can see pairs of intersecting lines in fabric patterns, at construction sites, and on road maps. If you know the measurement of one of these angles, you can find the measurements of the other three angles.

Complementary Angles

Two angles are complementary angles if and only if the sum of their measures is 90°. Each angle is a complement of the other. To be complementary, a pair of angles do not need to be adjacent.

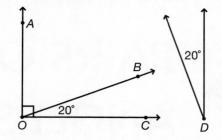

In the previous figure, ∠AOB and ∠BOC are a pair of complementary angles. ∠AOB and ∠D are also a pair of complementary angles. ∠AOB is a complement of ∠BOC and ∠D.

Practice

Find the measure of a complement of ∠1 for each of the following measures of ∠1.

_____ **1.** $m\angle 1 = 44°$

_____ **2.** $m\angle 1 = 68°$

_____ **3.** $m\angle 1 = 80°$

_____ **4.** $m\angle 1 = 25°$

_____ **5.** $m\angle 1 = 3°$

Supplementary Angles

Two angles are supplementary angles if and only if the sum of their measures is 180°. Each angle is a supplement of the other. Similar to complementary angles, supplementary angles do not need to be adjacent.

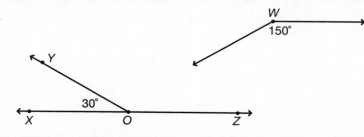

In this figure, ∠XOY and ∠YOZ are supplementary angles. ∠XOY and ∠W are also supplementary angles. ∠XOY is a supplement of ∠YOZ and ∠W.

Practice

Find the measure of a supplement of ∠2 for each of the following measures of ∠2.

_____ **6.** $m\angle 2 = 78°$

_____ **7.** $m\angle 2 = 130°$

_____ **8.** $m\angle 2 = 60°$

_____ **9.** $m\angle 2 = 155°$

_____ **10.** $m\angle 2 = 1°$

Vertical Angles

Vertical angles are two angles whose sides form two pairs of opposite rays. When two lines intersect, they form two pairs of vertical angles.

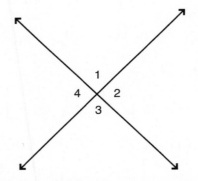

In the previous figure, the two pairs of vertical angles are ∠1 and ∠3, as well as ∠2 and ∠4. Also, ∠1 and ∠2 are supplementary angles. Since ∠1 and ∠3 are both supplements to the same angle, they are congruent; in other words, they have the same measurement.

> *Vertical Angles Theorem:* Vertical angles are congruent.

Practice

Use the following figure to answer practice problems 11–13.

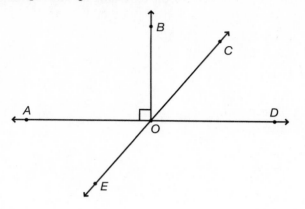

11. Name two supplement angles of ∠DOE.

12. Name a pair of complementary angles.

13. Name two pairs of vertical angles.

State whether the following statements are true or false.

———— **14.** Complementary angles must be acute.

———— **15.** Supplementary angles must be obtuse.

———— **16.** Two acute angles can be supplementary.

———— **17.** A pair of vertical angles can be complementary.

———— **18.** A pair of vertical angles can be supplementary.

———— **19.** Vertical angles must have the same measure.

———— **20.** Complementary angles can be adjacent.

———— **21.** Supplementary angles can be adjacent.

———— **22.** Any two right angles are supplementary.

———— **23.** Two acute angles are always complementary.

_____ **24.** An acute and an obtuse angle are always supplementary.

_____ **25.** The intersection of two rays creates two pairs of vertical angles and four pairs of supplementary angles.

Use the following figure to answer practice problem 26.

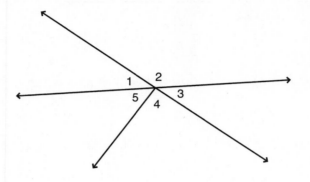

26. A common error is assuming that any pair of angles that are "across from each other" are vertical. In this figure, ∠1 and ∠3 are vertical angles because they are formed by intersecting lines. Angles 2 and 4 are not vertical angles. Name three other pairs of nonadjacent angles that are also not vertical.

Use the following figure to answer practice problems 27–30.

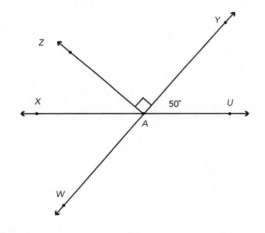

27. List the supplementary angles of ∠WAU.

28. List the complementary angles of ∠WAX.

29. List the vertical angles shown in the figure.

30. What is the measure of ∠ZAX.

Skill Building until Next Time

Find a pair of intersecting lines. You might use the lines in a floor tile, a fabric design, or a piece of furniture, or you could draw your own. Measure the four angles that the intersecting line forms to confirm the vertical angles theorem.

 6 ▶ TYPES OF TRIANGLES

Lesson Summary

In this lesson, you will learn how to classify triangles according to the lengths of their sides and their angle measurements.

I t would be difficult to name an occupation where classifying triangles is a required skill; however, it is a skill that will help you solve complex geometry problems. Each of the triangles discussed in this lesson has special properties that will help you solve problems.

Classification by Sides

You can classify triangles by the lengths of their sides. On the next page are three examples of special triangles called *equilateral*, *isosceles*, and *scalene*.

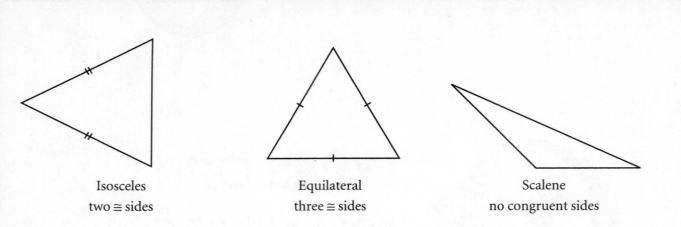

Isosceles
two ≅ sides

Equilateral
three ≅ sides

Scalene
no congruent sides

To show that two or more sides of a triangle have the same measurement, a hatch mark is made through the congruent sides. Sometimes, two hatch marks are made on each congruent side, and sometimes, three hatch marks are made on each congruent side. You can match up the number of hatch marks to find which sides are congruent. You'll see these hatch marks in most geometry books. The symbol for congruent is ≅.

Practice

Classify each triangle shown or described as equilateral, isosceles, or scalene.

1.

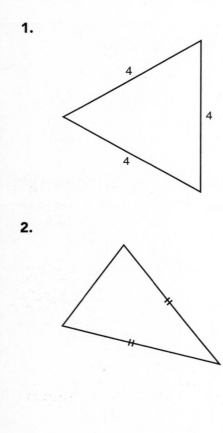

2.

3.

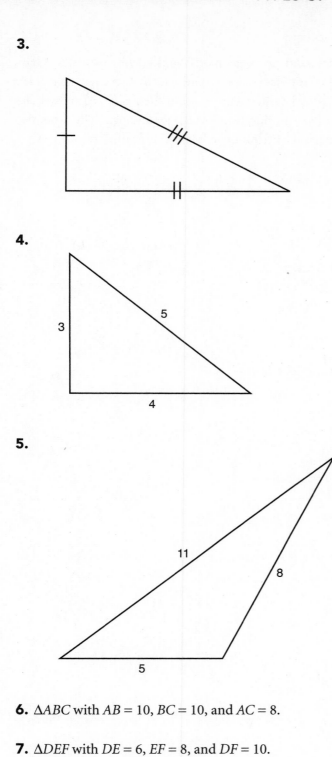

4.

5.

6. $\triangle ABC$ with $AB = 10$, $BC = 10$, and $AC = 8$.

7. $\triangle DEF$ with $DE = 6$, $EF = 8$, and $DF = 10$.

8. $\triangle XYZ$ with $XY = 7$, $YZ = 7$, and $XZ = 7$.

Isosceles Triangles

Isosceles triangles are important geometric figures to understand. Some geometry books define isosceles as having *at least* two congruent sides. For our purposes, we will define isosceles as having exactly two congruent sides. Did you know that the parts of an isosceles triangle have special names? The two congruent sides of an isosceles triangle are called the *legs*. The angle formed by the two congruent sides is called the *vertex angle*. The other two angles are called the *base angles*. And finally, the side opposite the vertex angle is called the *base*.

Example:

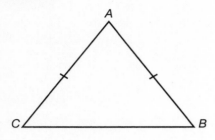

legs: $\overline{AC}$ and $\overline{AB}$ base angles: $\angle B$ and $\angle C$

vertex angle: $\angle A$ base: $\overline{BC}$

Practice

Name the legs, vertex angle, base angles, and base of the isosceles triangles.

9.

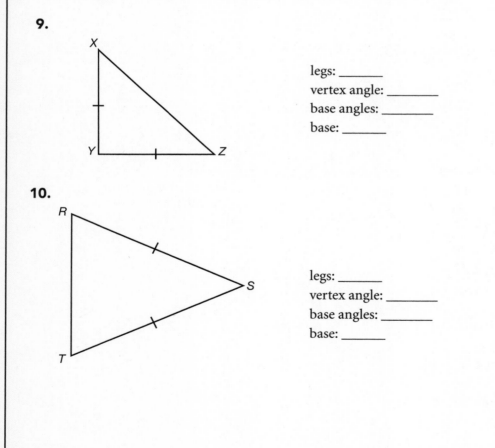

legs: _____

vertex angle: _____

base angles: _____

base: _____

10.

legs: _____

vertex angle: _____

base angles: _____

base: _____

11.

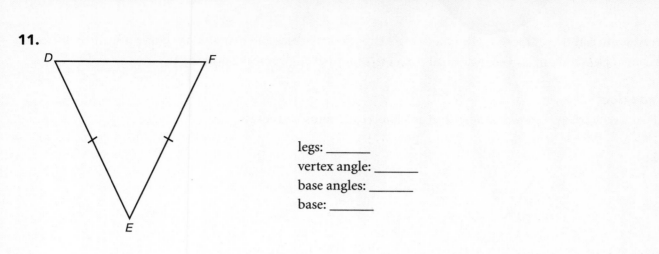

legs: _____
vertex angle: _____
base angles: _____
base: _____

Classification by Angles

You can also classify triangles by the measurements of their angles. Here are four examples of special triangles. They are called *acute, equiangular, right,* and *obtuse.*

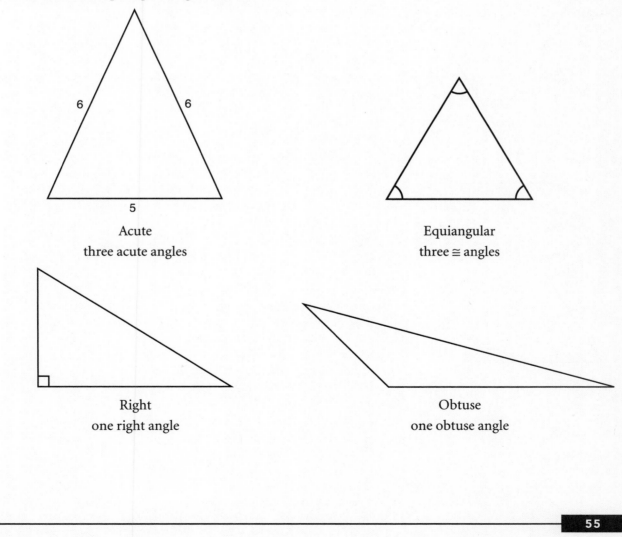

Acute
three acute angles

Equiangular
three ≅ angles

Right
one right angle

Obtuse
one obtuse angle

To show that two or more angles of a triangle have the same measurement, a small curve is made in the congruent angles. You can also use two small curves to show that angles are congruent.

Practice

Classify each triangle shown or described as acute, right, obtuse, or equiangular.

12.

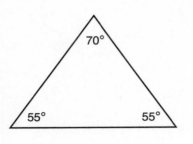

13.

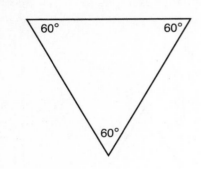

14.

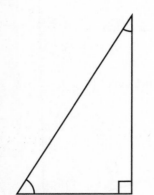

15.

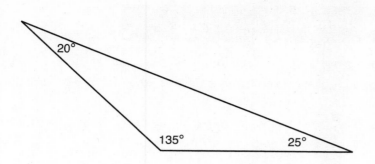

16. $\triangle MNO$ with $m\angle M = 130°$, $m\angle N = 30°$, and $m\angle O = 20°$.

17. $\triangle RST$ with $m\angle R = 80°$, $m\angle S = 45°$, and $m\angle T = 55°$.

18. $\triangle GHI$ with $m\angle G = 20°$, $m\angle H = 70°$, and $m\angle I = 90°$.

Determine whether the following statements are true or false.

_____ **19.** The vertex angle of an isosceles triangle will always be the smallest angle.

_____ **20.** An equilateral triangle is also an equiangular triangle.

_____ **21.** An isosceles triangle may also be a right triangle.

_____ **22.** It is possible for a scalene triangle to be equiangular.

_____ **23.** It is possible for a right triangle to be equilateral.

_____ **24.** It is possible for an obtuse triangle to be isosceles.

_____ **25.** An equilateral triangle is also an acute triangle.

_____ **26.** A triangle can have two obtuse angles.

_____ **27.** A triangle cannot have more than one right angle.

Give the answer to each question based on the readings from the chapter.

28. The non-congruent side of an isosceles triangle is called the _____.

29. List the types of triangles with at least two acute angles.

30. A small curve made on an angle of a triangle signifies _____.

Skill Building until Next Time

A triangle is called a rigid figure because the length of at least one side of a triangle must be changed before its shape changes. This is an important quality in construction. Notice the triangles you can see in floor and roof trusses.

To prove the strength of a triangle, place two glasses about four inches apart. Bridge the cups with a dollar bill.

Then try to balance a box of toothpicks on the dollar bill without the box of toothpicks touching the glasses.

Fold the dollar bill like a fan.

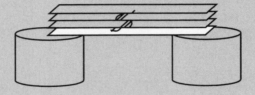

Now try to balance the box of toothpicks on the folded dollar bill.

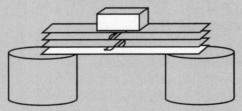

When you look at the folded dollar bill from the side, what shape do you see?

The view of the dollar bill from the side shows several triangles.

7 ▶ CONGRUENT TRIANGLES

Lesson Summary

This lesson will help you identify corresponding parts of congruent triangles and to name the postulate or theorem that shows that two triangles are congruent.

Congruent triangles are commonly used in the construction of quilts, buildings, and bridges. Congruent triangles are also used to estimate inaccessible distances, such as the width of a river or the distance across a canyon. In this lesson, you will learn simple ways to determine whether two triangles are congruent.

When you buy floor tiles, you get tiles that are all the same shape and size. One tile will fit right on top of another. In geometry, you would say one tile is *congruent* to another tile. Similarly, in the following figure, $\triangle ABC$ and $\triangle XYZ$ are congruent. They have the same size and shape.

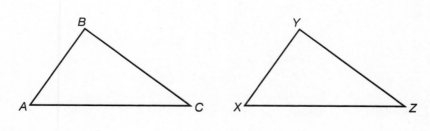

Imagine sliding one triangle over to fit on top of the other triangle. You would put point A on point X; point B on point Y; and point C on point Z. When the vertices are matched in this way, $\angle A$ and $\angle X$ are called *corresponding angles*, and $\overline{AB}$ and $\overline{XY}$ are called *corresponding sides*.

Corresponding angles and corresponding sides are often referred to as corresponding parts of the triangles. In other words, you could say <u>C</u>orresponding <u>P</u>arts of <u>C</u>ongruent <u>T</u>riangles are <u>C</u>ongruent. This statement is often referred to by the initials CPCTC.

When $\triangle ABC$ is congruent to $\triangle XYZ$, you write $\triangle ABC \cong \triangle XYZ$. This means that all of the following are true:

$$\angle A \cong \angle X \qquad \angle B \cong \angle Y \qquad \angle C \cong \angle Z$$
$$\overline{AB} \cong \overline{XY} \qquad \overline{BC} \cong \overline{YZ} \qquad \overline{AC} \cong \overline{XZ}$$

Suppose instead of writing $\triangle ABC \cong \triangle XYZ$, you started to write $\triangle CAB \cong \underline{\hspace{1cm}}$. Since you started with C to name the first triangle, you must start with the corresponding letter, Z, to name the second triangle. Corresponding parts are named in the same order. If you name the first triangle $\triangle CAB$, then the second triangle must be named $\triangle ZXY$. In other words, $\triangle CAB \cong \triangle ZXY$.

Example: Name the corresponding angles and corresponding sides.

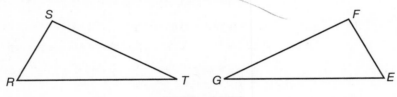

$$\triangle RST \cong \triangle EFG$$

Solution:
Corresponding angles: $\angle R$ and $\angle E$; $\angle S$ and $\angle F$; $\angle T$ and $\angle G$
Corresponding sides: $\overline{RS}$ and $\overline{EF}$; $\overline{ST}$ and $\overline{FG}$; $\overline{RT}$ and $\overline{EG}$

Practice

For practice problems 1–6, complete each statement, given $\triangle JKM \cong \triangle PQR$.

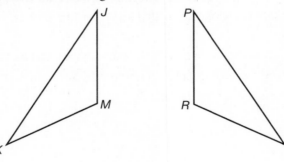

1. $\angle M$ corresponds to $\angle$_____.

2. $\angle P$ corresponds to $\angle$_____.

3. $\angle Q$ corresponds to $\angle$_____.

4. $\overline{JK}$ corresponds to _____.

5. $\overline{RQ}$ corresponds to _____.

6. $\overline{PR}$ corresponds to _____.

For practice problems 7–10, complete each statement, given $\triangle GFH \cong \triangle JFH$.

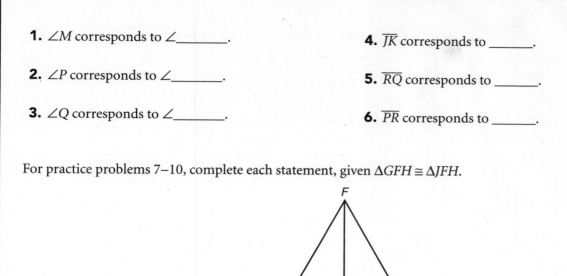

7. $\triangle FGH \cong \triangle$_____

8. $\triangle HJF \cong \triangle$_____

9. $\triangle FHJ \cong \triangle$_____

10. $\triangle JFH \cong \triangle$_____

Side-Side-Side (SSS) Postulate

If you have three sticks that make a triangle and a friend has identical sticks, would it be possible for each of you to make different-looking triangles? No, it is impossible to do this. A postulate of geometry states this same idea. It is called the *side-side-side postulate*.

> *Side-Side-Side Postulate:* If three sides of one triangle are congruent to three sides of another triangle, then the two triangles are congruent.

Take a look at the following triangles to see this postulate in action:

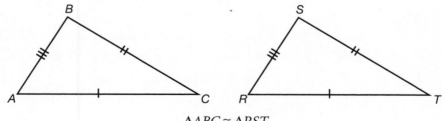

$$\triangle ABC \cong \triangle RST$$

The hatch marks on the triangles show which sides are congruent to which in the two triangles. For example, $\overline{AC}$ and $\overline{RT}$ both have one hatch mark, which shows that these two segments are congruent. $\overline{BC}$ is congruent to $\overline{ST}$, as shown by the two hatch marks, and $\overline{AB}$ and $\overline{RS}$ are congruent as shown by the three hatch marks.

Since the markings indicate that the three pairs of sides are congruent, you can conclude that the three pairs of angles are also congruent. From the definition of congruent triangles, it follows that all six parts of $\triangle ABC$ are congruent to the corresponding parts of $\triangle RST$.

Practice

Use the following figure to answer questions 11–15.

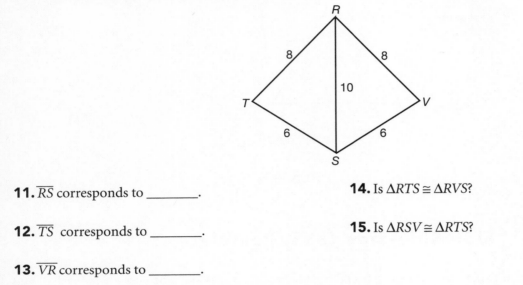

11. $\overline{RS}$ corresponds to _____.

12. $\overline{TS}$ corresponds to _____.

13. $\overline{VR}$ corresponds to _____.

14. Is $\triangle RTS \cong \triangle RVS$?

15. Is $\triangle RSV \cong \triangle RTS$?

Side-Angle-Side (SAS) Postulate

If you put two sticks together at a certain angle, there is only one way to finish forming a triangle. Would it be possible for a friend to form a different-looking triangle if he or she started with the same two lengths and the same angle? No, it would be impossible. Another postulate of geometry states this same idea; it is called the *side-angle-side postulate*.

> *Side-Angle-Side Postulate:* If two sides and the included angle of one triangle are congruent to the corresponding parts of another triangle, then the triangles are congruent.

Look at the following two triangles to see an example of this postulate:

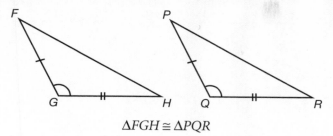

$$\triangle FGH \cong \triangle PQR$$

Practice

Use the following figure to answer practice problems 16–20.

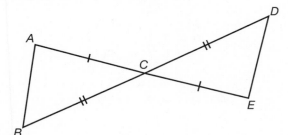

16. What kind of angles are $\angle ACB$ and $\angle ECD$?

17. Is $\angle ACB \cong \angle ECD$?

18. $\overline{CE}$ corresponds to _____.

19. $\overline{BC}$ corresponds to _____.

20. Is $\triangle ACB \cong \triangle EDC$?

Angle-Side-Angle (ASA) Postulate

There is one more postulate that describes two congruent triangles. *Angle-side-angle* involves two angles and a side between them. The side is called an included side.

> *Angle-Side-Angle Postulate:* If two angles and the included side of one triangle are congruent to corresponding parts of another triangle, then the triangles are congruent.

Take a look at the following two triangles:

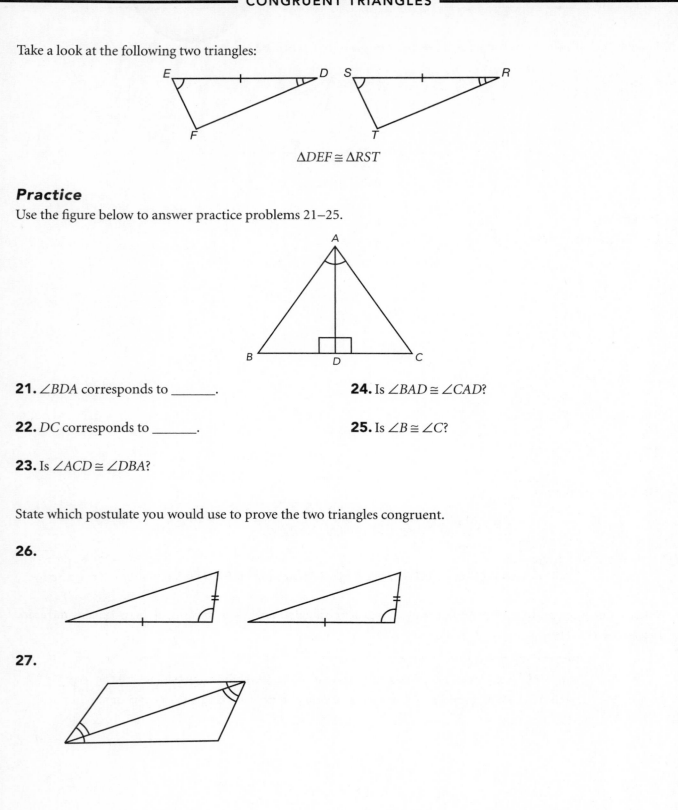

$$\Delta DEF \cong \Delta RST$$

Practice

Use the figure below to answer practice problems 21–25.

21. $\angle BDA$ corresponds to _____.

22. DC corresponds to _____.

23. Is $\angle ACD \cong \angle DBA$?

24. Is $\angle BAD \cong \angle CAD$?

25. Is $\angle B \cong \angle C$?

State which postulate you would use to prove the two triangles congruent.

26.

27.

28.

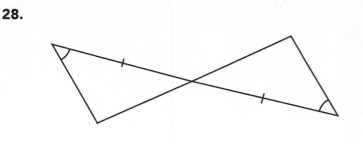

29.

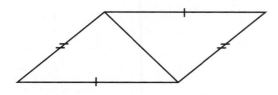

30.

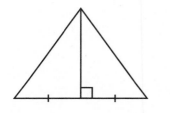

Skill Building until Next Time

A carpenter's square is a tool used in carpentry. Similar tools are used for arts and crafts projects, as well as for wallpaper hanging. With the recent craze for home improvement, you may either have a carpenter's square or may be able to borrow one from a neighbor. A carpenter's square can be used to bisect an angle at the corner of a board. Get a carpenter's square and mark equal lengths $\overline{WX}$ and $\overline{WZ}$ along the edges. Put the carpenter's square on a board so that $\overline{XY} = \overline{YZ}$. Mark point Y and draw $\overline{WY}$.

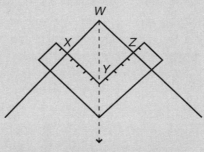

a) Which postulate can you use to show that $\triangle WXY \cong \triangle WZY$?

b) Why is $\angle XWY \cong \angle ZWY$?

8 ▶ TRIANGLES AND THE PYTHAGOREAN THEOREM

Lesson Summary

In this lesson, you will learn the special names for the sides of a right triangle. You will also learn how to use the Pythagorean theorem to find missing parts of a right triangle and to determine whether three segments will make a right triangle.

The right triangle is very important in geometry because it can be used in so many different ways. The Pythagorean theorem is just one of the special relationships that can be used to help solve problems and find missing information. Right triangles can be used to find solutions to problems involving figures that are not even polygons.

Parts of a Right Triangle

In a right triangle, the sides that meet to form the right angle are called the *legs*. The side opposite the right angle is called the *hypotenuse*. The hypotenuse is always the longest of the three sides. It is important that you can correctly identify the sides of a right triangle, regardless of what position the triangle is in.

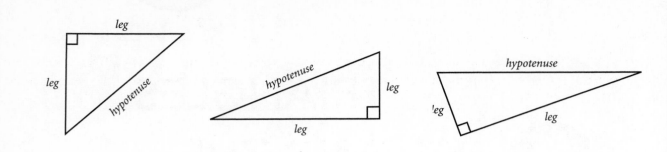

Practice

ΔABC is a right triangle.

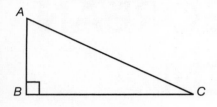

1. Name the legs.

2. Name the hypotenuse.

ΔPQR is a right triangle.

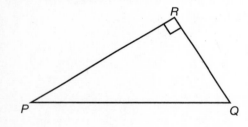

3. Name the legs.

4. Name the hypotenuse.

ΔHLM is a right triangle.

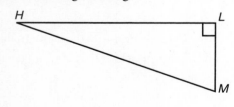

5. Name the legs.

6. Name the hypotenuse.

Review of Squares and Square Roots

Before you study the Pythagorean theorem, let's first review squares and square roots. Just like addition and subtraction are inverses, so are squares and square roots. In other words, they "undo" each other. To square a number, you multiply it by itself. For example, 5^2 means two factors of five, or five times five, which is 25. Written algebraically, it looks like this: $5^2 = 5 \times 5 = 25$. A common mistake is to say that you multiply by two, since two is the exponent (the small raised number). But the exponent tells you how many times to use the base (bottom number) as a factor.

Twenty-five is a perfect square. It can be written as the product of two equal factors. It would be helpful for you to learn the first 16 perfect squares. When completed, the following chart will be a useful reference. It is not necessarily important that you memorize the chart, but you need to understand how the numbers are generated. Even the most basic calculators can help you determine squares and square roots of larger numbers.

Practice

7. Complete the chart.

NUMBER	SQUARE
1	1
2	4
3	9
4	
5	25
	36
7	49
8	
	81
10	100
11	
12	144
	169
14	196
15	
16	

The Pythagorean Theorem

The Pythagorean theorem is one of the most famous theorems in mathematics. The Greek mathematician Pythagoras (circa 585–500 B.C.) is given credit for originating it. Evidence shows that it was used by the Egyptians and Babylonians for hundreds of years before Pythagoras.

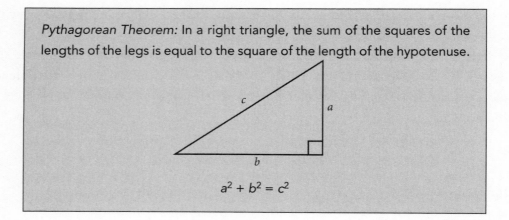

Pythagorean Theorem: In a right triangle, the sum of the squares of the lengths of the legs is equal to the square of the length of the hypotenuse.

$$a^2 + b^2 = c^2$$

The Pythagorean theorem can be used to solve many real-life problems. Any unknown length can be found if you can make it a part of a right triangle. You need to know only two of the sides of a right triangle to find the third unknown side. A common mistake is always adding the squares of the two known lengths. You add the squares of the legs only when you are looking for the hypotenuse. If you know the hypotenuse and one of the legs, then you subtract the square of the leg from the square of the hypotenuse. Another common mistake is forgetting to take the square root as your final step. You just need to remember that you are solving not for the square of the side, but for the length of the side.

Examples: Find each missing length.

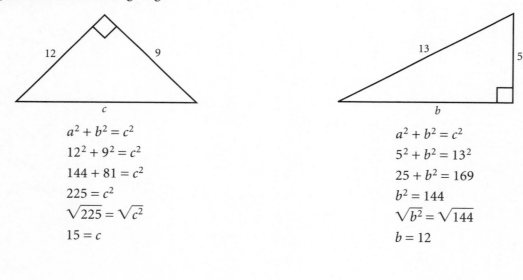

$$a^2 + b^2 = c^2$$
$$12^2 + 9^2 = c^2$$
$$144 + 81 = c^2$$
$$225 = c^2$$
$$\sqrt{225} = \sqrt{c^2}$$
$$15 = c$$

$$a^2 + b^2 = c^2$$
$$5^2 + b^2 = 13^2$$
$$25 + b^2 = 169$$
$$b^2 = 144$$
$$\sqrt{b^2} = \sqrt{144}$$
$$b = 12$$

Practice

Find each missing length.

8.

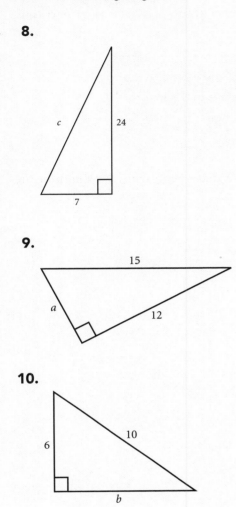

9.

10.

Use this parallelogram to answer questions 11 and 12.

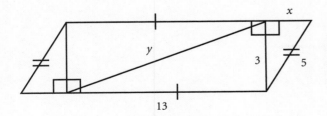

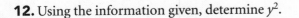

11. Determine x.

12. Using the information given, determine y^2.

Converse of the Pythagorean Theorem

Another practical use of the Pythagorean theorem involves determining whether a triangle is acute, right, or obtuse. This involves using the converse of the Pythagorean theorem.

> *Converse of the Pythagorean Theorem:* If the square of the length of the longest side of a triangle is equal to the sum of the squares of the lengths of the two shorter sides, then the triangle is a right triangle.

Examples: Determine whether the following are right triangles. Note that c represents the length of the longest side in each triangle.

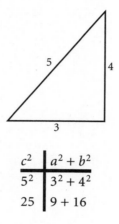

c^2	$a^2 + b^2$
5^2	$3^2 + 4^2$
25	9 + 16

Since 25 = 9 + 16, the three sides with lengths of 5, 3, and 4 make a right triangle.

Here is another example:

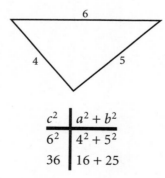

c^2	$a^2 + b^2$
6^2	$4^2 + 5^2$
36	16 + 25

Since 36 ≠ 16 + 25, the three sides with lengths of 6, 4, and 5 do not make a right triangle.

Practice

The lengths of three sides of a triangle are given. Determine whether the triangles are right triangles.

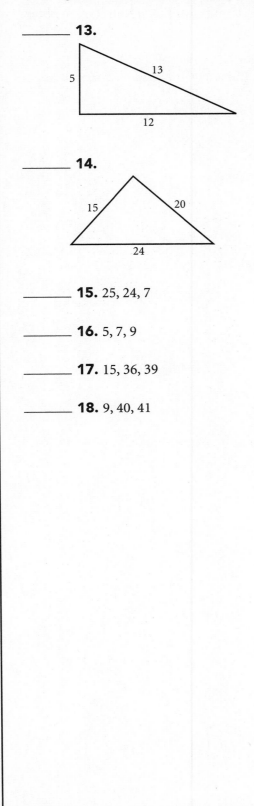

_____ **13.**

_____ **14.**

_____ **15.** 25, 24, 7

_____ **16.** 5, 7, 9

_____ **17.** 15, 36, 39

_____ **18.** 9, 40, 41

Acute and Obtuse Triangles

If you have determined that a triangle is not a right triangle, then you can determine whether it is an acute or obtuse triangle by using one of the following theorems:

Theorem: If the square of the length of the longest side is greater than the sum of the squares of the lengths of the other two shorter sides, then the triangle is obtuse.

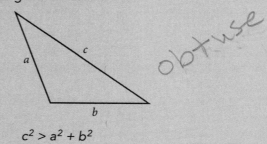

$$c^2 > a^2 + b^2$$

Theorem: If the square of the length of the longest side is less than the sum of the squares of the lengths of the other two shorter sides, then the triangle is acute.

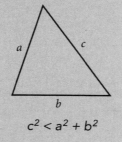

$$c^2 < a^2 + b^2$$

Let's take a look at a couple of examples to see these theorems in action.

Example: In $\triangle ABC$, the lengths of the sides are $c = 24$, $a = 15$, $b = 20$.

c^2	$a^2 + b^2$
24^2	$15^2 + 20^2$
576	225 + 400
576	625

Since $576 < 625$, this is an acute triangle.

Example: In $\triangle ABC$, the lengths of the sides are $c = 9$, $a = 5$, $b = 7$.

c^2	$a^2 + b^2$
9	$5^2 + 7^2$
81	25 + 49
81	74

Since $81 > 74$, this is an obtuse triangle.

Practice

The lengths of the three sides of a triangle are given. Classify each triangle as acute, right, or obtuse.

19. 30, 40, 50   *obt*

20. 10, 11, 13

21. 2, 10, 11

22. 7, 7, 10   *obt*

23. 50, 14, 28

24. 5, 6, 7

25. 8, 12, 7

Julie drew the following star, and wanted to know more about the properties of some of the triangles she created in the process. Use the Pythagorean theorem and the concepts of the lesson to answer questions 26 and 27.

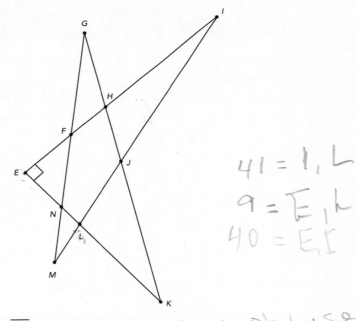

$$41 = IL$$
$$9 = EL$$
$$40 = EI$$

26. $\overline{IL}$ = 41 cm and $\overline{EL}$ = 9 cm. Find $\overline{EI}$ and determine the type of triangle.   *Obtuse*

27. $\overline{EK}$ = 36 ft. and $\overline{EH}$ = 27 ft. Find $\overline{HK}$ and determine the type of triangle.

Use the following figure to answer question 28–30.

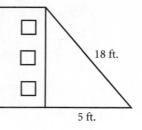

18 ft.

5 ft.

28. How far up a building will an 18-foot ladder reach if the ladder's base is 5 feet from the building? Express your answer to the nearest foot. Solve the problem when the ladder is 3 feet from the building. Why would it be impractical to solve the problem if the base of the ladder was closer than 3 feet from the building?

29. If instead of an 18-foot ladder, a 25-foot ladder was used, how far would the ladder need to be from the base of the building in order for it to reach a window that is 20 feet from the ground?

30. If a ladder is going to be placed 6 feet from the base of the wall and needs to reach a window that is 8 feet from the ground, how long must the ladder be?

Skill Building until Next Time

Look around your home for examples of right triangles. How many can you find? Measure the three sides of the triangle to make sure it is a right triangle.

9 ▶ PROPERTIES OF POLYGONS

Lesson Summary

In this lesson, you will learn how to determine whether a figure is a polygon. You will also learn how to identify concave and convex polygons. You will learn how to classify polygons by their sides and how to find the measures of their interior and exterior angles.

The word *polygon* comes from Greek words meaning "many angled." A polygon is a closed plane figure formed by line segments. The line segments are called *sides* that intersect only at their endpoints, which are called *vertex points*.

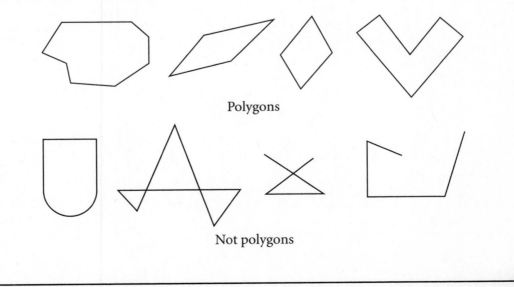

Polygons

Not polygons

A polygon is convex when no segment connecting two vertices (vertex points) contains points outside the polygon. In other words, if you wrapped a rubber band around a convex polygon, it would fit snugly without gaps. A concave polygon has at least one place that "caves in."

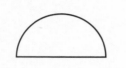

Convex polygons   Concave polygons

Practice

Classify each figure as a convex polygon, a concave polygon, or not a polygon.

1.

2.

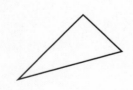

3.

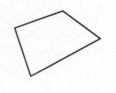

4.

5.

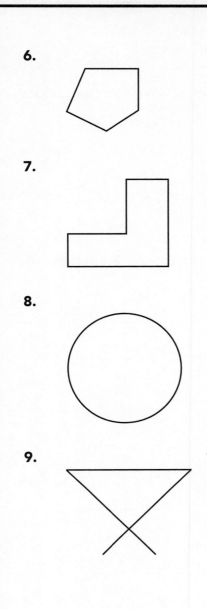

6.

7.

8.

9.

Parts of a Polygon

Two sides of a polygon that intersect are called *consecutive* or *adjacent sides*. The endpoints of a side are called *consecutive* or *adjacent vertices*. The segment that connects two nonconsecutive vertices is called a *diagonal* of the polygon.

Naming Polygons

When naming a polygon, you name its consecutive vertices in clockwise or counterclockwise order. Here are a few of the ways to name the following polygon: *ABCDE*, *DEABC*, or *EDCBA*.

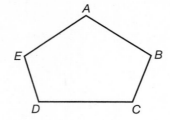

Although you can start at any vertex, you cannot skip around when you name a polygon. For example, you cannot name the polygon *BDEAC*, *ECBAD*, or *ACEDB*.

Polygons are classified by their number of sides.

NUMBER OF SIDES	POLYGON
3	triangle
4	quadrilateral
5	pentagon
6	hexagon
7	heptagon
8	octagon
9	nonagon
10	decagon

Practice

Classify each polygon by the number of sides, and determine if each is correctly named.

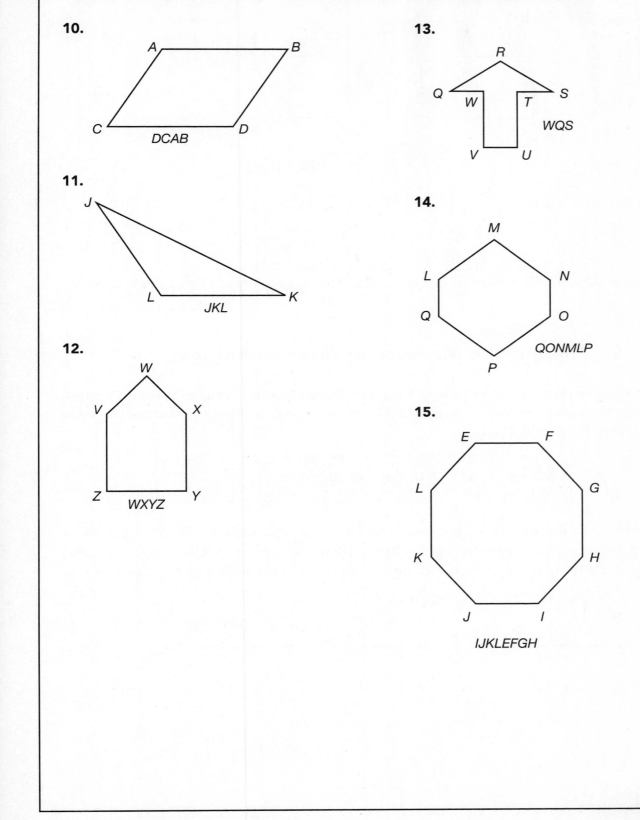

10.

A   B

C   D

DCAB

11.

J

L   K

JKL

12.

W

V   X

Z   Y

WXYZ

13.

R

Q   S
W   T

V   U

WQS

14.

M

L   N

Q   O

P

QONMLP

15.

E   F

L   G

K   H

J   I

IJKLEFGH

Which of the following are acceptable names for each polygon?

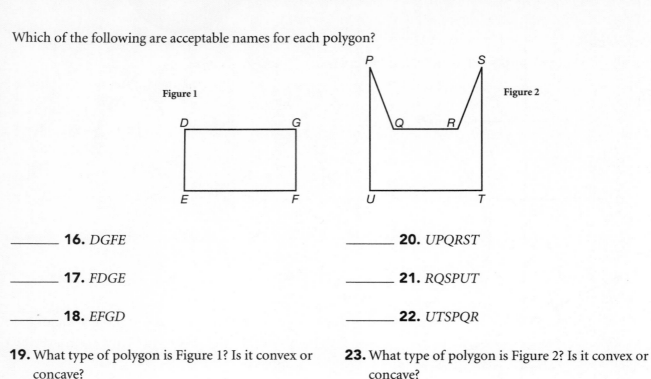

Figure 1

Figure 2

_____ **16.** *DGFE*

_____ **17.** *FDGE*

_____ **18.** *EFGD*

_____ **20.** *UPQRST*

_____ **21.** *RQSPUT*

_____ **22.** *UTSPQR*

19. What type of polygon is Figure 1? Is it convex or concave?

23. What type of polygon is Figure 2? Is it convex or concave?

Finding the Measure of Interior Angles

There are two theorems that you can use to find the measure of interior angles of a convex polygon. One theorem works only for triangles. The other theorem works for all convex polygons, including triangles. Let's take a look at the theorem for triangles first.

> *Theorem:* The sum of the interior angles of a triangle is 180°.

To illustrate this, cut a triangle from a piece of paper. Tear off the three angles or points of the triangle. Without overlapping the edges, put the vertex points together. They will form a straight line or straight angle. Remember that a straight angle is 180°; therefore, the three angles of a triangle add up to 180°, as shown in the following figures.

You can find the sum of the interior angles of a convex polygon if you know how many sides the polygon has. Look at these figures. Do you see a pattern?

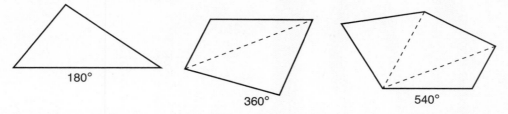

The diagram suggests that polygons can be divided into triangles. Since each triangle has 180°, multiply the number of triangles by 180 to get the sum of the interior angles.

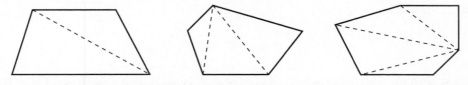

Look for a pattern in the number of sides a polygon has and the number of triangles drawn from one vertex point. You will always have two fewer triangles than the number of sides of the polygon. You can write this as a general statement with the letter *n* representing the number of sides of the polygon.

Theorem: If a convex polygon has *n* sides, then its angle sum is given by this formula:

$$S = 180(n - 2)$$

Example: Find the sum of the interior angles of a polygon with 12 sides.

Solution: $n = 12$

$S = 180(n - 2)$

$S = 180(12 - 2)$

$S = 180(10)$

$S = 1,800$

Therefore, the sum of the interior angles of a 12-sided polygon is 1,800°.

Practice

Find the sum of the interior angles for each figure.

24.   **25.**

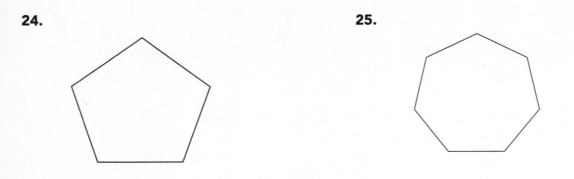

Finding the Measure of Interior Angles of Regular Polygons

A *regular polygon* is any polygon whose interior angles all have the same angle measurement. Recall that the formula to calculate the sum, S, of the interior angles of a polygon with n sides is $S = 180(n-2)$. In order to calculate the measure of the interior angle of a regular polygon, divide the sum, S, by the number of sides, n.

> *Theorem:* The measure of an interior angle, x, of a regular polygon with n sides is given by the formula:
> $$\frac{180(n-2)}{n}$$

Example: Find the measure of an interior angle of a regular pentagon.

Solution: $n = 5$

$$x = \frac{180(n-2)}{n}$$

$$x = \frac{180(5-2)}{5}$$

$$x = \frac{180(3)}{5}$$

$$x = 108$$

Each interior angle in a pentagon will measure 108°.

Finding the Measure of Exterior Angles

Use this theorem to find the measure of exterior angles of a convex polygon.

> *Theorem*: The sum of the exterior angles of a convex polygon is always 360°.

To illustrate this theorem, picture yourself walking alongside a polygon. As you reach each vertex point, you will turn the number of degrees in the exterior angle. When you return to your starting point, you will have rotated 360°.

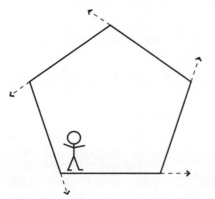

This figure shows this theorem using a pentagon. Do you see that this would be true for all polygons as stated in the theorem?

Practice

26. Complete the table for convex polygons.

NUMBER OF SIDES	6	10	14	16
Interior ∠ sum				
Exterior ∠ sum				

27. The home plate used in baseball and softball has this shape:

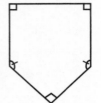

As you can see, there are three right angles and two congruent obtuse angles. What are the measures of the two obtuse angles?

28. Find the measure of an interior angle of a regular octagon.

29. Find the measure of an interior angle of a regular decagon.

30. Find the measure of an *exterior* angle of a regular hexagon. (Hint: Recall the sum of the exterior angles of a convex polygon.)

Skill Building until Next Time

Observe your surroundings today and see how many examples of convex and concave polygons you can find.

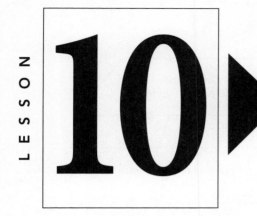

10 ▶ QUADRILATERALS

Lesson Summary

In this lesson, you will learn how to name and classify special quadrilaterals. You will also learn how to use the special properties associated with parallelograms, rectangles, rhombuses, squares, and trapezoids.

Quadrilaterals are one of the most commonly used figures in buildings, architecture, and design. The diagram on the next page shows the characteristics and relationships among the special quadrilaterals.

All four-sided polygons are classified as quadrilaterals. *Quadrilaterals* branch off into two distinctive subgroups: parallelograms and trapezoids. Trapezoids are quadrilaterals that have only one pair of opposite parallel sides. If the trapezoid has congruent legs, then the figure is an isosceles trapezoid. The diagram on page 88 shows that an isosceles trapezoid is one type of trapezoid, which is one type of quadrilateral. In other words, the figures become more specialized as the chart flows downward.

The other main branch of quadrilaterals consists of parallelograms. Parallelograms have two pairs of opposite parallel sides. Parallelograms branch off into two special categories: rectangles and rhombuses. A rectangle is a parallelogram with four congruent angles. A rhombus is a parallelogram with four congruent sides. A square is a parallelogram with four congruent angles and four congruent sides. In other words, a square is also a rectangle and a rhombus.

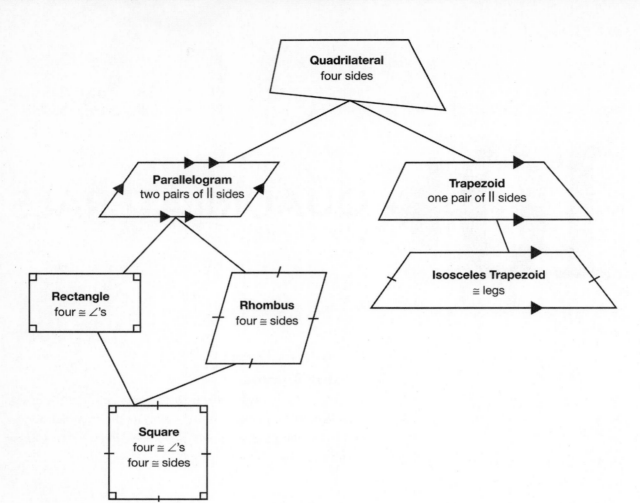

Practice

Use the diagram to determine whether each statement is true or false.

_____ **1.** All trapezoids are quadrilaterals.

_____ **2.** All parallelograms are rectangles.

_____ **3.** All rectangles are quadrilaterals.

_____ **4.** All parallelograms are squares.

_____ **5.** All squares are both rectangles and rhombuses.

_____ **6.** All rectangles are both parallelograms and quadrilaterals.

_____ **7.** All rhombuses are squares.

_____ **8.** All isosceles trapezoids are parallelograms.

_____ **9.** All squares are trapezoids.

_____ **10.** All quadrilaterals are squares.

Properties of Parallelograms

The following properties of parallelograms will help you determine if a figure is a parallelogram or just a quadrilateral. The properties are also useful to determine measurements of angles, sides, and diagonals of parallelograms.

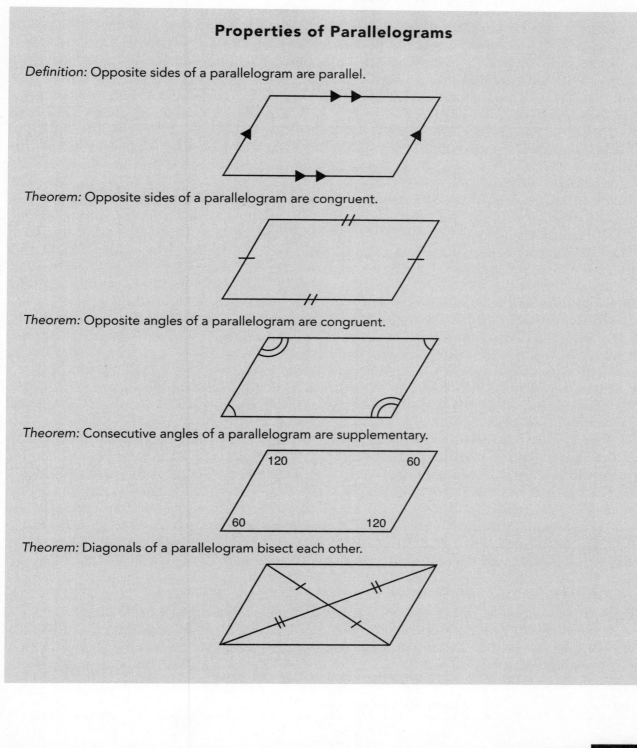

Properties of Parallelograms

Definition: Opposite sides of a parallelogram are parallel.

Theorem: Opposite sides of a parallelogram are congruent.

Theorem: Opposite angles of a parallelogram are congruent.

Theorem: Consecutive angles of a parallelogram are supplementary.

Theorem: Diagonals of a parallelogram bisect each other.

Note that diagonals of a parallelogram are not necessarily congruent. Watch out for this, because it is a common error.

Examples: *BMAH* is a parallelogram.

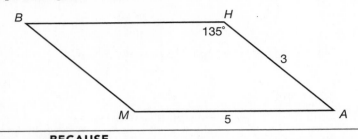

WE KNOW THAT	BECAUSE
$BM = 3$	Opposite sides are congruent.
$BH = 5$	Opposite sides are congruent.
$m\angle M = 135°$	Opposite angles are congruent.
$m\angle A = 45°$	Consecutive angles are supplementary.
$m\angle B = 45°$	Opposite angles are congruent.

WXYZ is a parallelogram.

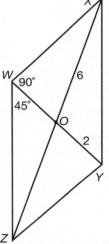

WE KNOW THAT	BECAUSE
$m\angle XWZ = 90° + 45° = 135°$	Angle addition postulate
$m\angle XYZ = 135°$	Opposite angles are congruent.
$m\angle WXY = 45°$	Consecutive angles are supplementary.
$m\angle WZY = 45°$	Opposite angles are congruent.
$WO = 2$	Diagonals bisect each other.
$ZO = 6$	Diagonals bisect each other.

Practice

Use the following figure of parallelogram *ABCD* to answer questions 11–15.

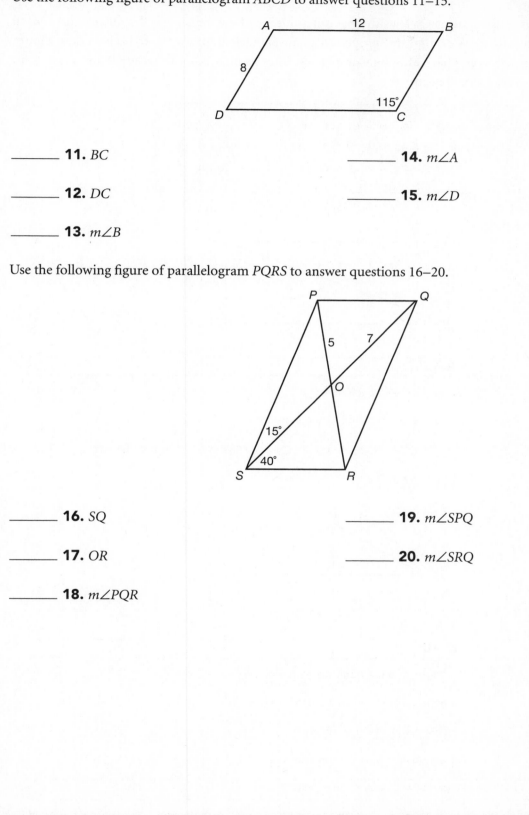

_____ **11.** *BC*

_____ **12.** *DC*

_____ **13.** *m∠B*

_____ **14.** *m∠A*

_____ **15.** *m∠D*

Use the following figure of parallelogram *PQRS* to answer questions 16–20.

_____ **16.** *SQ*

_____ **17.** *OR*

_____ **18.** *m∠PQR*

_____ **19.** *m∠SPQ*

_____ **20.** *m∠SRQ*

Other Special Properties

The rectangle, rhombus, and square have a few other special properties. First, remember that these figures are all parallelograms; therefore, they possess the same properties as any parallelogram. However, because these figures are special parallelograms, they also have additional properties. Since a square is both a rectangle and a rhombus, a square possesses these same special properties.

> *Theorem:* The diagonals of a rectangle are congruent.
>
> *Theorem:* The diagonals of a rhombus are perpendicular, and they bisect the angles of the rhombus.

Examples: *NEWS* is a rectangle.

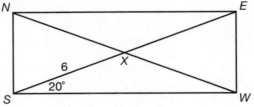

WE KNOW THAT	BECAUSE
$XE = 6$	Diagonals bisect each other.
$NW = 12$	Diagonals of a rectangle are congruent.
$\angle NES = 20°$	Alternate interior angles are congruent, when formed by parallel lines.
$\angle NSW = 90°$	Definition of a rectangle.

ABCD is a rhombus.

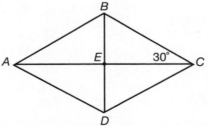

WE KNOW THAT	BECAUSE
$m\angle BEC = 90°$	Diagonals of a rhombus are perpendicular.
$m\angle DCE = 30°$	Diagonals of a rhombus bisect the angles.
$m\angle BAD = 60°$	Opposite angles are congruent.
$m\angle ABC = 120°$	Consecutive angles are supplementary.
$m\angle ADC = 120°$	Opposite angles are congruent.

Practice

Use the following figure to find the side length and angle measures for rectangle *PQRS* for questions 21–24.

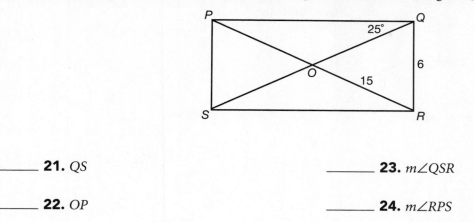

_____ **21.** *QS*

_____ **22.** *OP*

_____ **23.** *m∠QSR*

_____ **24.** *m∠RPS*

Use the Pythagorean theorem to solve questions 25 and 26.

25. Find *PQ*, given *OR* = 5 and *QR* = 6.

26. Determine the perimeter of rectangle *PQRS*.

Use the following figure to find the angle measures for rhombus *GHJK* for questions 27–32.

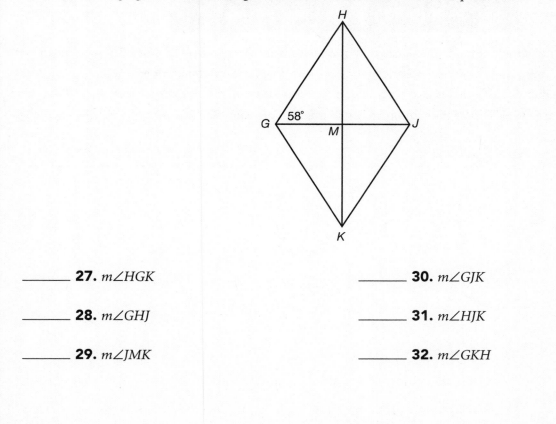

_____ **27.** *m∠HGK*

_____ **28.** *m∠GHJ*

_____ **29.** *m∠JMK*

_____ **30.** *m∠GJK*

_____ **31.** *m∠HJK*

_____ **32.** *m∠GKH*

Use the following diagram to answer question 33.

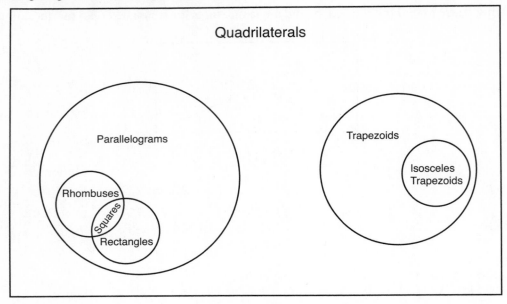

33. Explain how this Venn diagram and the flowchart shown at the beginning of this lesson show the relationships among quadrilaterals.

Skill Building until Next Time

Look around your environment to find examples of quadrilaterals and make a list of them. Then go down your list and see how many correct geometric names you can give for each item. For example, the state of Tennessee is a quadrilateral that is also a parallelogram. Your desktop may be a rectangle, a parallelogram, and a quadrilateral. The Venn diagram shown in practice problem 33 may be useful to you.

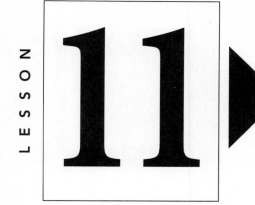

11 ▶ RATIO, PROPORTION, AND SIMILARITY

Lesson Summary

In this lesson, you will learn how to write and simplify ratios. You will also learn how to determine whether two ratios are a proportion and how to use proportions to solve problems. In addition, you will learn how to determine whether two triangles are similar.

Ratios and proportions have many applications. Architects use them when they make scale models of buildings. Interior designers use scale drawings of rooms to decide furniture size and placement. Similar triangles can be used to find indirect measurements. Measurements of distances such as heights of tall buildings and the widths of large bodies of water can be found using similar triangles and proportions. Let's begin by looking at ratios.

What Is a Ratio?

If you compare two quantities, then you have used a ratio. A *ratio* is the comparison of two numbers using division. The ratio of x to y can be written $\frac{x}{y}$ or $x{:}y$. Ratios are usually expressed in simplest form.

Examples: Express each ratio in simplest form.

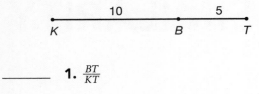

(a) $\frac{DR}{RY}$ (b) $\frac{RY}{DR}$ (c) $\frac{DR}{DY}$

Solutions:

(a) $\frac{4}{12} = \frac{1}{3}$ (b) $\frac{12}{4} = 3$ (c) $\frac{4}{16} = \frac{1}{4}$

Practice

Express each ratio in simplest form.

K —— 10 —— B —— 5 —— T

_____ **1.** $\frac{BT}{KT}$

2. Write the ratio $\frac{KB}{BT}$ another way.

_____ **3.** $\frac{BT}{KB}$

4. Write the ratio $\frac{BT}{KB}$ another way.

_____ **5.** $\frac{BT}{KT}$

6. Write the ratio $\frac{BT}{KT}$ another way.

What Is a Proportion?

Since $\frac{2}{4}$ and $\frac{3}{6}$ are both equal to $\frac{1}{2}$, they are equal to each other. A statement that two ratios are equal is called a *proportion*. A proportion can be written in one of the following ways:

$\frac{2}{4} = \frac{3}{6}$ or   2:4 = 3:6

The first and last numbers in a proportion are called the *extremes*. The middle numbers are called the *means*.

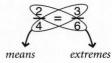

means   extremes

Means-Extremes Property

In a proportion, the product of the means equals the product of the extremes.

If $\frac{a}{b} = \frac{c}{d}$, then $ad = bc$.

If $a{:}b = c{:}d$, then $ad = bc$.

Examples: Tell whether each of the following is a proportion.

(a) $\frac{3}{6} = \frac{1}{2}$ (b) $2{:}5 = 4{:}10$ (c) $\frac{1}{3} = \frac{2}{9}$

Solutions:

(a) $3 \times 2 = 6 \times 1$ (b) $2 \times 10 = 5 \times 4$ (c) $1 \times 9 = 3 \times 2$

     $6 = 6$ $20 = 20$ $9 \neq 6$

     yes   yes   no

Practice

Determine whether each of the following is a proportion.

_____ **7.** $\frac{2}{7} = \frac{4}{14}$

8. State the means for the proportion $\frac{2}{7} = \frac{4}{14}$.

_____ **9.** $12{:}16 = 9{:}15$

10. State the extremes for the proportion $12{:}16 = 9{:}15$.

_____ **11.** $2{:}3 = 3{:}2$

12. State the means for the proportion $2{:}3 = 3{:}2$.

Solving Proportion Problems

Proportions can also be used to solve problems. When three parts of a proportion are known, you can find the fourth part by using the *means-extremes property*.

Examples: Solve each proportion.

(a) $\frac{4}{x} = \frac{2}{10}$ (b) $\frac{a}{7} = \frac{12}{21}$ (c) $\frac{5}{2} = \frac{15}{y}$

Solutions:

(a) $2x = 4 \times 10$ (b) $21a = 7 \times 12$ (c) $5y = 2 \times 15$

     $2x = 40$ $21a = 84$ $5y = 30$

     $x = 20$ $a = 4$ $y = 6$

Practice

Solve each proportion.

_____ **13.** $\frac{x}{8} = \frac{1}{2}$

_____ **14.** $\frac{8}{y} = \frac{2}{11}$

_____ **15.** $\frac{2}{7} = \frac{8}{z}$

_____ **16.** $\frac{2}{5} = \frac{a}{20}$

_____ **17.** $\frac{8}{b} = \frac{4}{7}$

_____ **18.** $\frac{9}{x} = \frac{36}{4}$

Triangle Similarity

You can prove that two figures are similar by using the definition of *similar*. Two figures are similar if you can show that the following two statements are true:

(1) Corresponding angles are congruent.

(2) Corresponding sides are in proportion.

In addition to using the definition of similar, you can use three other methods for proving that two triangles are similar: the angle-angle postulate, the side-side-side postulate, and the side-angle-side postulate.

If you know the measurements of two angles of a triangle, can you find the measurement of the third angle? Yes, from Lesson 9, you know that the sum of the three angles of a triangle is 180°. Therefore, if two angles of one triangle are congruent to two angles of another triangle, then their third angles must also be congruent. This will help you understand the next postulate. You should know that the symbol used for similarity is ~.

> *Angle-Angle Postulate (AA Postulate):* If two angles of one triangle are congruent to two angles of another triangle, then the triangles are similar.

Examples: Are these triangles similar?

(a)

Solutions:

(a) $\angle A \cong \angle D$, given
$\angle BCA \cong \angle ECD$, vertical $\angle$'s are $\cong$
$\triangle ABC \sim \triangle DEC$, AA postulate

(b)

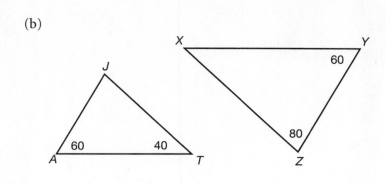

(b) $m\angle J = 180 - (60 + 40)$
    $m\angle J = 80$
    $\triangle AJT \sim \triangle YZX$, AA postulate

Practice

Determine whether each pair of triangles is similar.

_____ **19.**

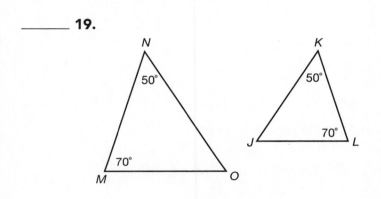

_____ **20.**

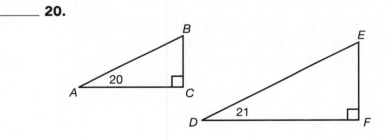

_____ **21.**

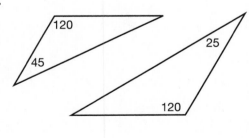

_____ **22.**

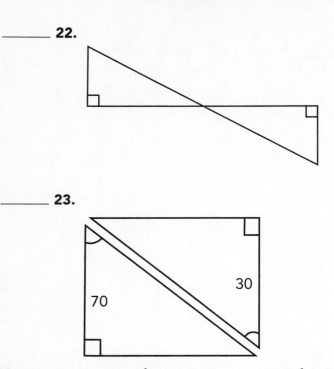

_____ **23.**

Here are two more postulates you can use to prove that two triangles are similar:

> _Side-Side-Side Postulate (SSS Postulate):_ If the lengths of the corresponding sides of two triangles are proportional, then the triangles are similar.
> _Side-Angle-Side postulate (SAS postulate):_ If the lengths of two pairs of corresponding sides of two triangles are proportional and the corresponding included angles are congruent, then the triangles are similar.

Examples: Which postulate, if any, could you use to prove that each pair of triangles is similar?

(a)

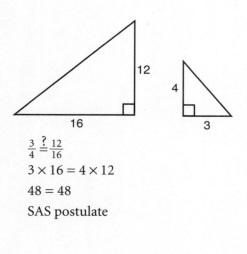

$\dfrac{3}{4} \overset{?}{=} \dfrac{12}{16}$

$3 \times 16 = 4 \times 12$

$48 = 48$

SAS postulate

(b)

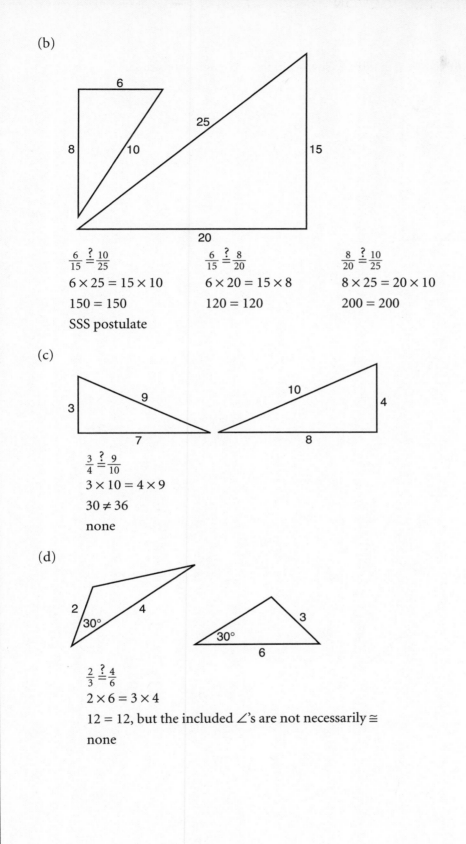

$\frac{6}{15} \overset{?}{=} \frac{10}{25}$

$6 \times 25 = 15 \times 10$

$150 = 150$

SSS postulate

$\frac{6}{15} \overset{?}{=} \frac{8}{20}$

$6 \times 20 = 15 \times 8$

$120 = 120$

$\frac{8}{20} \overset{?}{=} \frac{10}{25}$

$8 \times 25 = 20 \times 10$

$200 = 200$

(c)

$\frac{3}{4} \overset{?}{=} \frac{9}{10}$

$3 \times 10 = 4 \times 9$

$30 \neq 36$

(d)

$\frac{2}{3} \overset{?}{=} \frac{4}{6}$

$2 \times 6 = 3 \times 4$

$12 = 12$, but the included $\angle$'s are not necessarily $\cong$

Practice

Which postulate, if any, could you use to prove that the following pairs of triangles are similar?

24.

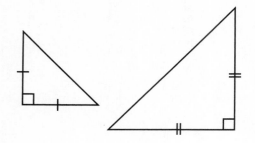

25.

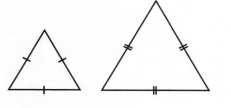

26.

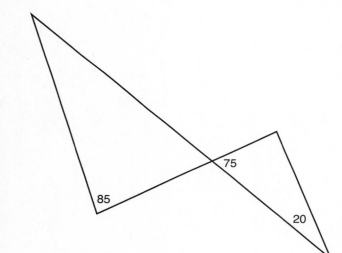

27.

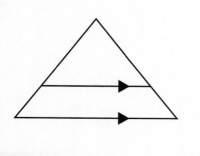

28.

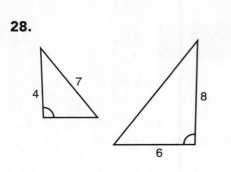

29.

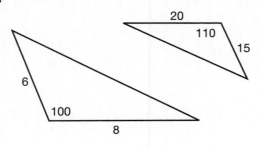

30.

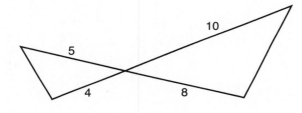

31.

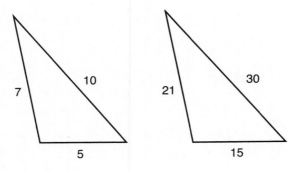

Skill Building until Next Time

Outside your school, workplace, or home, find two objects that are fairly close to each other. One should be short enough to measure with a measuring tape. Some good choices would be a stop sign or a firehydrant. The other should be too tall to measure with a measuring tape. A flagpole or a lightpost would be good choices. Measure the length of the shadows of the short and the tall objects. Set up a proportion:

$$\frac{\text{height of the flagpole}}{\text{shadow of the flagpole}} = \frac{\text{height of the stop sign}}{\text{shadow of the stop sign}}$$

Replace the height of the flagpole with x and the other parts of the proportion with the measurements that you found. Solve by using the *means-extremes property.*

12 ▶ PERIMETER OF POLYGONS

Lesson Summary

In this lesson, you will learn how to find the distance around convex and concave polygons. You will also learn how to use formulas to find the perimeter of polygons.

Perimeter is found by measuring the distance around an object. Crown molding around a room and a fence around a garden are just two examples of where you might need to find a perimeter. To find the perimeter of an irregular shape, simply add all the lengths of its sides. You can find the perimeters of polygons, which have a uniform shape, by using formulas. Carpenters and landscape architects use perimeters on a regular basis.

Examples: Find the perimeter of each polygon.

(a)   (b)

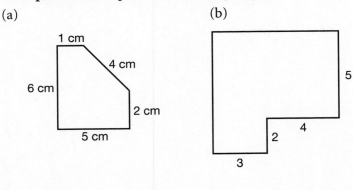

Solutions:

(a)   perimeter = 1 + 4 + 2 + 5 + 6

          = 18 cm

(b)

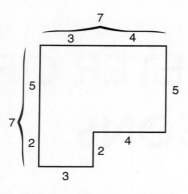
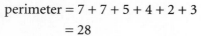

perimeter = 7 + 7 + 5 + 4 + 2 + 3

          = 28

Practice

Find the perimeter of each polygon.

_____ **1.**

_____ **2.**

_____ **3.**

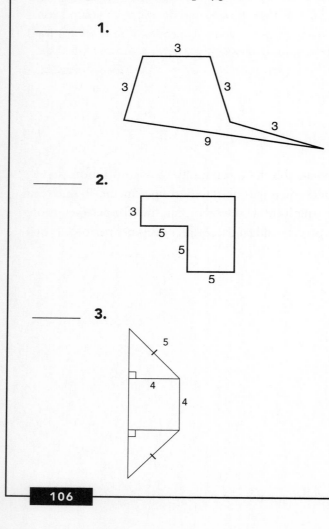

_____ **4.**

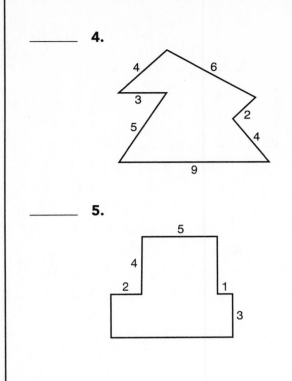

_____ **5.**

Perimeter Formulas

For certain polygons, you can find the perimeter in a more efficient manner. Using a standard formula for a particular shape is faster and easier than adding all the sides. Here are the most commonly used perimeter formulas:

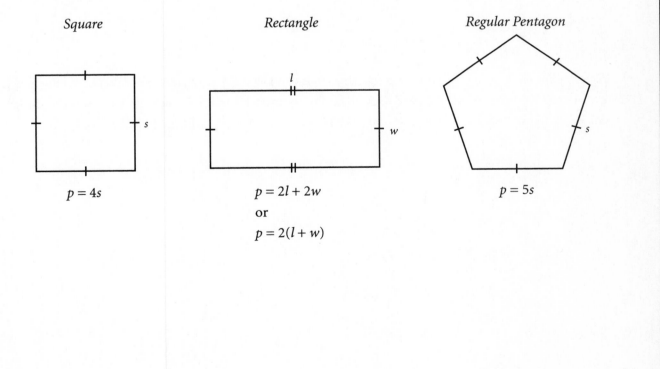

Square

$p = 4s$

Rectangle

$p = 2l + 2w$
or
$p = 2(l + w)$

Regular Pentagon

$p = 5s$

Examples: Use a formula to find each perimeter.

(a)

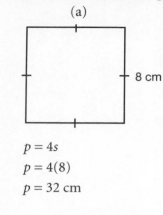

8 cm

$p = 4s$
$p = 4(8)$
$p = 32$ cm

(b)

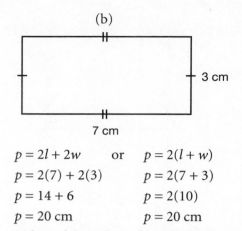

3 cm

7 cm

$p = 2l + 2w$ or $p = 2(l + w)$
$p = 2(7) + 2(3)$ $p = 2(7 + 3)$
$p = 14 + 6$ $p = 2(10)$
$p = 20$ cm $p = 20$ cm

Either of these formulas will always work.
You can choose which one is easier for you.

(c)

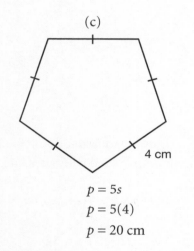

4 cm

$p = 5s$
$p = 5(4)$
$p = 20$ cm

A *regular polygon* is a polygon with all congruent angles and all congruent sides. So if you know the length of one side of a regular polygon, all you need to do to find its perimeter is multiply the number of sides by the length of a side. In other words, $p = ns$, where n = number of sides and s = the length of each side.

Practice

Use a formula to find the perimeter of each polygon.

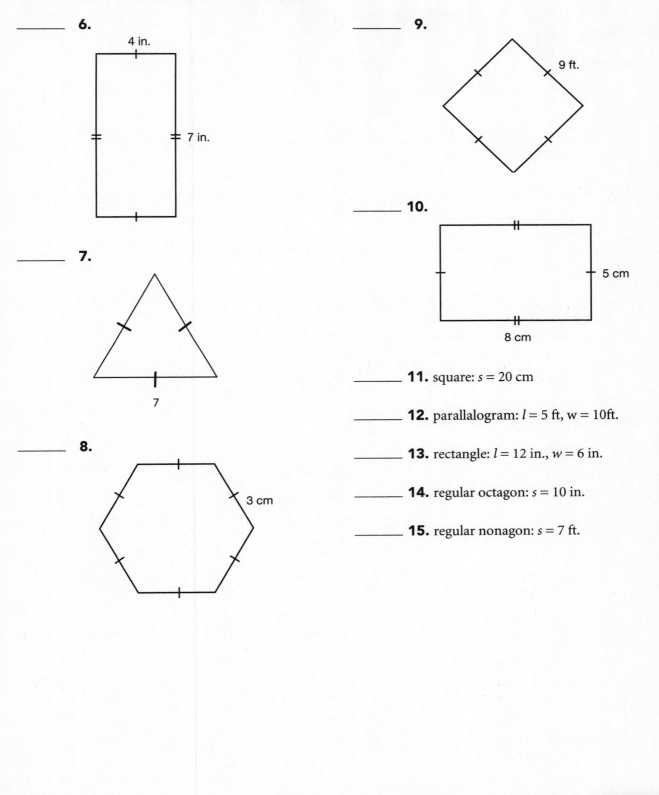

_____ **6.**

4 in.

7 in.

_____ **7.**

7

_____ **8.**

3 cm

_____ **9.**

9 ft.

_____ **10.**

5 cm

8 cm

_____ **11.** square: $s = 20$ cm

_____ **12.** parallalogram: $l = 5$ ft, w = 10ft.

_____ **13.** rectangle: $l = 12$ in., $w = 6$ in.

_____ **14.** regular octagon: $s = 10$ in.

_____ **15.** regular nonagon: $s = 7$ ft.

Working Backward

If you know the perimeter of a polygon, you can also use a perimeter formula to find the length of one of its sides by working backward.

Examples:

(a)

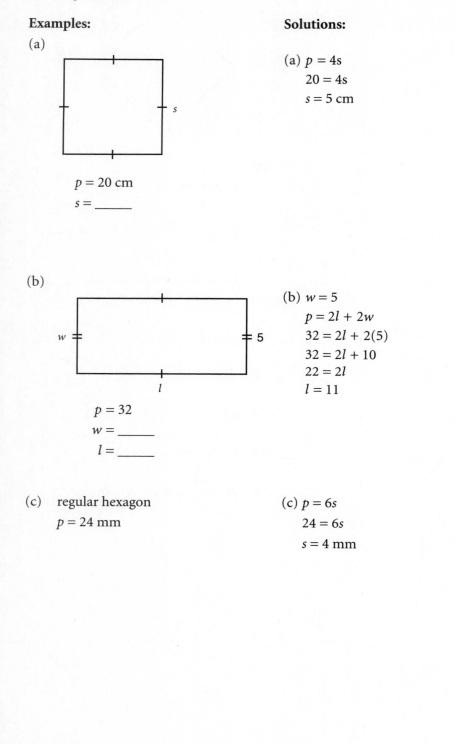

$p = 20$ cm

$s =$ _____

(b)

$p = 32$

$w =$ _____

$l =$ _____

(c)   regular hexagon

$p = 24$ mm

Solutions:

(a) $p = 4s$

$\quad 20 = 4s$

$\quad s = 5$ cm

(b) $w = 5$

$\quad p = 2l + 2w$

$\quad 32 = 2l + 2(5)$

$\quad 32 = 2l + 10$

$\quad 22 = 2l$

$\quad l = 11$

(c) $p = 6s$

$\quad 24 = 6s$

$\quad s = 4$ mm

Practice

Find the length of the indicated side(s).

16.

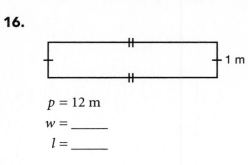

1 m

$p = 12$ m

$w =$ _____

$l =$ _____

17.

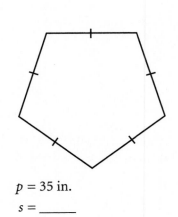

10 cm

$p = 30$ cm

$w =$ _____

$l =$ _____

18.

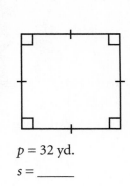

$p = 35$ in.

$s =$ _____

19.

$p = 32$ yd.

$s =$ _____

20. equilateral triangle: $p = 60$ cm $\quad s =$ _____

21. regular octagon: $p = 80$ cm $\quad s =$ _____

22. regular hexagon: $p = 48$ cm $\quad s =$ _____

23. regular decagon: $p = 230$ cm $\quad s =$ _____

24. rhombus: $p = 44$ cm $\quad s =$ _____

Find answers to these practical problems.

25. Suppose you want to frame an 8-inch by 10-inch picture. How much molding will you need to buy?

26. A fence must be placed around your vegetable garden. The dimensions of the garden are 30 feet by 10 feet. How much fencing will you need?

27. You have a square dining room that you would like to trim with crown molding. If the length of the room is 17 feet, how much crown molding should you purchase?

28. Suppose you want to hang holiday icicle lights around your house. If your house is 32 feet by 40 feet and one package of lights is 9 feet long, how many packages of lights should you buy?

29. Suppose a bookshelf has three shelves that are each 3 feet long by 1 foot tall. You have books that are all 2 in. wide by 12 in. tall. How many of your books will fit in the bookcase?

30. You have a wall that is 9 ft. tall that you want to cover with posters. The posters you like each measure 3 ft. by 3 ft. If you bought 15 posters and placed them next to each other from floor to ceiling on the wall, how wide would the portion of the wall covered posters be?

Skill Building until Next Time

Use a tape measure to find the perimeter of your room, the door of your room, and a window in your room. Use formulas when possible.

LESSON

13 ▶ AREA OF POLYGONS

Lesson Summary

In this lesson, you will learn how to use formulas to find the areas of rectangles, parallelograms, triangles, and trapezoids. You will also learn how to use the formulas to work backward to find missing lengths of polygons.

People often confuse area and perimeter. As you learned in Lesson 12, perimeter is the distance around an object. In this lesson, you'll work with *area*, which is the amount of surface covered by an object. For example, the number of tiles on a kitchen floor would be found by using an area formula, while the amount of baseboard used to surround the room would be found by using a perimeter formula. Perimeter is always expressed in linear units. Area is always expressed in square units. You can not measure the amount of surface that is covered by an object by simply measuring it in one direction, you need to measure the object in two directions, like a square. A flat surface has two dimensions: length and width. When you multiply a number by itself, the number is said to be squared. In the same way, when two units of measurement are multiplied by each other, as in area, the unit is expressed in square units. By looking at a tiled floor, it is easy to see that *area* refers to how many squares it takes to cover a surface.

Finding the Area of a Rectangle

For a rectangle, the base can be any side. The base length is represented by b. The sides perpendicular to the base are referred to as the height. The height is referred to as h. The base is often called the length, l, and the height is often called the width, w. Length, l, and width, w, are used in the same manner as base, b, and height, h. This book uses *base* and *height*.

Here is a useful theorem you can use to find the area of a rectangle.

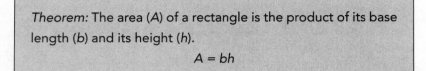

Theorem: The area (A) of a rectangle is the product of its base length (b) and its height (h).

$$A = bh$$

Examples: Find the area of each rectangle.

(a)

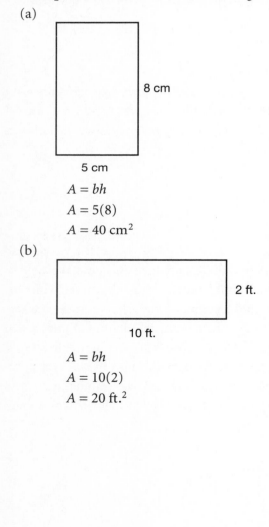

8 cm

5 cm

$A = bh$
$A = 5(8)$
$A = 40 \text{ cm}^2$

(b)

2 ft.

10 ft.

$A = bh$
$A = 10(2)$
$A = 20 \text{ ft.}^2$

Practice

Find the area of each rectangle.

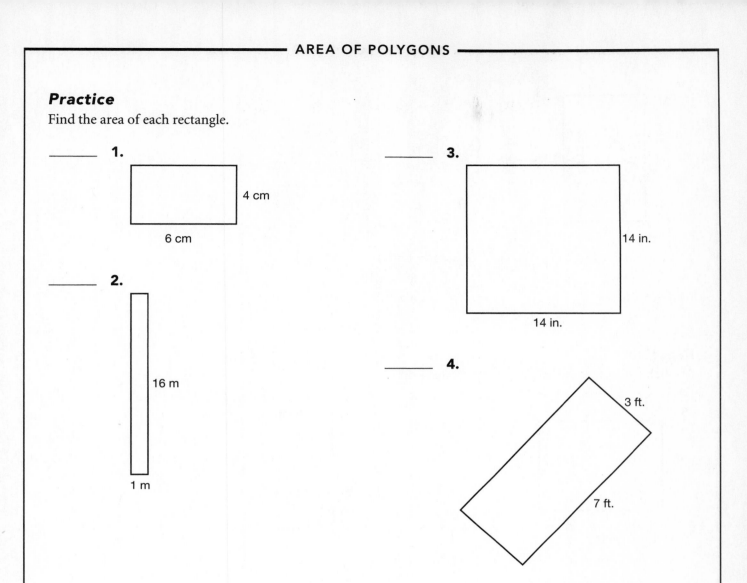

_____ **1.**

4 cm

6 cm

_____ **2.**

16 m

1 m

_____ **3.**

14 in.

14 in.

_____ **4.**

3 ft.

7 ft.

Finding the Length of an Unknown Side of a Rectangle

You can also use the area formula for a rectangle to find the length of an unknown side if you know the area.

Examples: For each rectangle, find the length of the indicated sides.

(a)

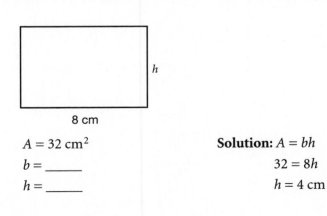

h

8 cm

$A = 32 \text{ cm}^2$

$b =$ _____

$h =$ _____

Solution: $A = bh$

$32 = 8h$

$h = 4 \text{ cm}$

(b)

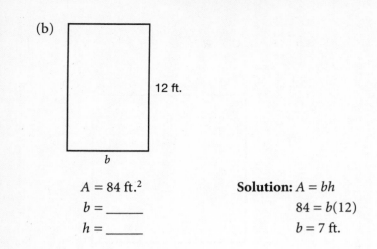

12 ft.

b

$A = 84$ ft.^{2}

$b = \underline{\hspace{1cm}}$

$h = \underline{\hspace{1cm}}$

Solution: $A = bh$

$84 = b(12)$

$b = 7$ ft.

Practice

For each figure, find the length of the indicated side(s).

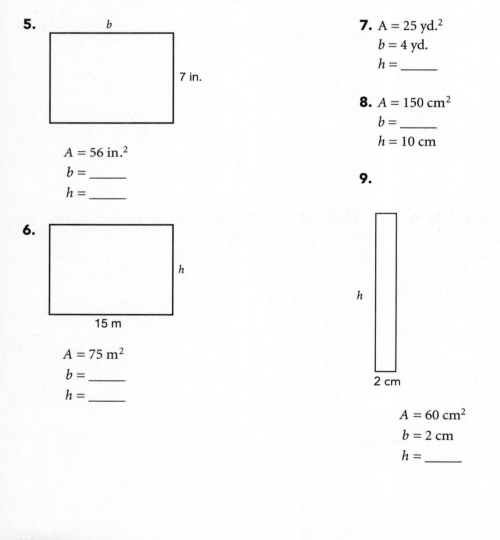

5.

b

7 in.

$A = 56$ in.^{2}

$b = \underline{\hspace{1cm}}$

$h = \underline{\hspace{1cm}}$

6.

h

15 m

$A = 75$ m^2

$b = \underline{\hspace{1cm}}$

$h = \underline{\hspace{1cm}}$

7. A $= 25$ yd.^{2}

$b = 4$ yd.

$h = \underline{\hspace{1cm}}$

8. $A = 150$ cm^2

$b = \underline{\hspace{1cm}}$

$h = 10$ cm

9.

h

2 cm

$A = 60$ cm^2

$b = 2$ cm

$h = \underline{\hspace{1cm}}$

Finding the Area and Unknown Sides of a Parallelogram

Any side of a parallelogram can be called the base. The height is the length of the altitude. The *altitude* is a segment perpendicular to the base.

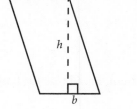 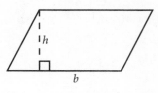

Draw an altitude of a parallelogram. Cut along the altitude to separate the parallelogram into two pieces. Fit the two pieces together to form a rectangle. You'll find that the base and height of the rectangle coincide with the base and altitude of the parallelogram.

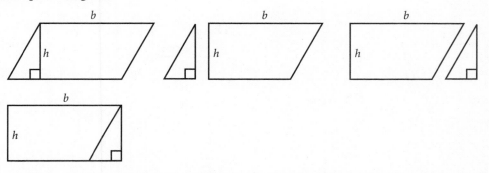

Using this information, can you predict the area formula for a parallelogram? Take a moment to make your prediction, then look at the following theorem.

> *Theorem:* The area (*A*) of a parallelogram is the product of its base length (*b*) and its height (*h*).
>
> $$A = bh$$

Examples: Find the area of the parallelogram.

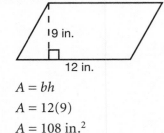

9 in.

12 in.

Solution: $A = bh$

$A = 12(9)$

$A = 108 \text{ in.}^2$

Find the indicated lengths.

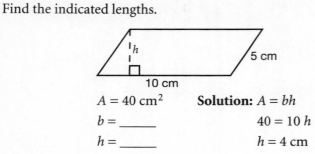

$A = 40 \text{ cm}^2$ **Solution:** $A = bh$

$b =$ _____ $40 = 10\,h$

$h =$ _____ $h = 4 \text{ cm}$

Note that the 5 cm measurement is unnecessary information for this problem. Recall that the base and height must be perpendicular to each other.

Practice

Find the indicated information.

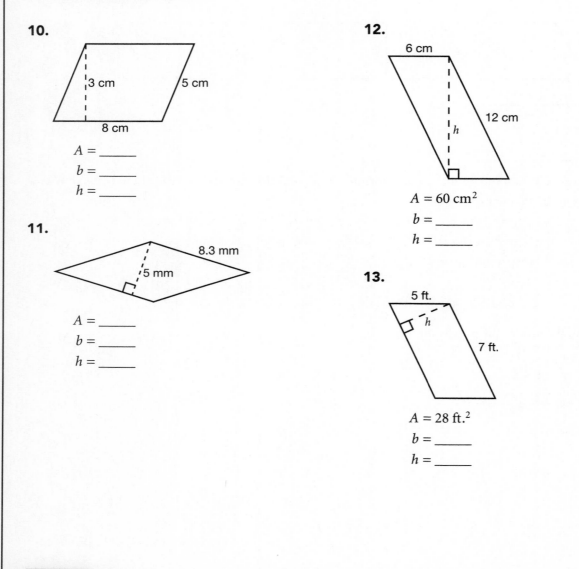

10.

$A =$ _____

$b =$ _____

$h =$ _____

11.

$A =$ _____

$b =$ _____

$h =$ _____

12.

$A = 60 \text{ cm}^2$

$b =$ _____

$h =$ _____

13.

$A = 28 \text{ ft.}^2$

$b =$ _____

$h =$ _____

Finding the Area and Unknown Sides of a Triangle

Look at the following figures and try to predict the area formula for a triangle.

Now see if your prediction is correct by reading the theorem below.

> *Theorem:* The area (A) of any triangle is half the product of its base length (b) and height (h).
> $$A = \frac{1}{2}bh$$

Examples: Find the area.

(a)

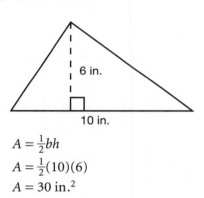

$A = \frac{1}{2}bh$

$A = \frac{1}{2}(10)(6)$

$A = 30 \text{ in.}^2$

(b)

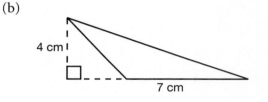

$A = \frac{1}{2}bh$

$A = \frac{1}{2}(7)(4)$

$A = 14 \text{ cm}^2$

Practice

Find the area of each triangle.

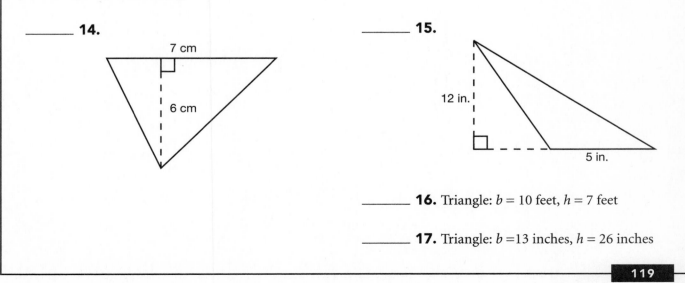

_____ **14.**

7 cm

6 cm

_____ **15.**

12 in.

5 in.

_____ **16.** Triangle: $b = 10$ feet, $h = 7$ feet

_____ **17.** Triangle: $b = 13$ inches, $h = 26$ inches

18.

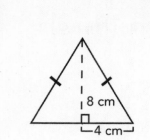

Find the indicated length.

19.

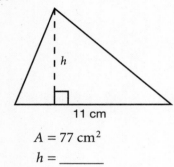

$A = 77 \text{ cm}^2$

$h = \underline{\hspace{1cm}}$

20.

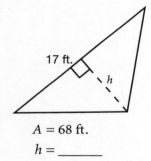

$A = 68 \text{ ft.}$

$h = \underline{\hspace{1cm}}$

21.

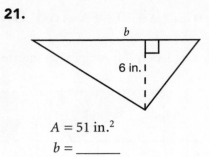

$A = 51 \text{ in.}^2$

$b = \underline{\hspace{1cm}}$

22. For a given triangle, A = 160 ft.^{2} with a base of 8 ft. What is the height?

23. For a given triangle, A = 64 in.^{2} with a height of 16 in. What is the base?

Finding the Area and Unknown Sides of a Trapezoid

Can you predict the area formula for a trapezoid? Look at these figures:

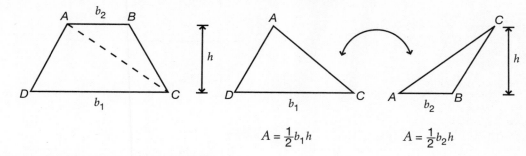

$$A = \tfrac{1}{2}b_1 h \qquad\qquad A = \tfrac{1}{2}b_2 h$$

Area of the trapezoid = area of two triangles
$$= \tfrac{1}{2}b_1 h + \tfrac{1}{2}b_2 h$$
$$= \tfrac{1}{2}h(b_1 + b_2)$$

> *Theorem:* The area of a trapezoid is half the product of the height and the sum of the base lengths $(b_1 + b_2)$.
> $$A = \tfrac{1}{2}h(b_1 + b_2)$$

Examples: Find the area or indicated length.

(a)

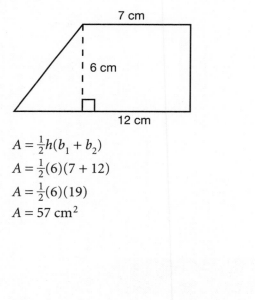

$$A = \tfrac{1}{2}h(b_1 + b_2)$$
$$A = \tfrac{1}{2}(6)(7 + 12)$$
$$A = \tfrac{1}{2}(6)(19)$$
$$A = 57 \text{ cm}^2$$

(b)

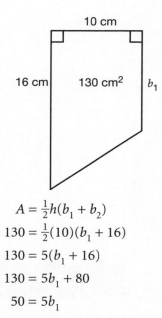

$$A = \tfrac{1}{2}h(b_1 + b_2)$$
$$130 = \tfrac{1}{2}(10)(b_1 + 16)$$
$$130 = 5(b_1 + 16)$$
$$130 = 5b_1 + 80$$
$$50 = 5b_1$$
$$b_1 = 10 \text{ cm}$$

Practice

Find the area or indicated length.

24.

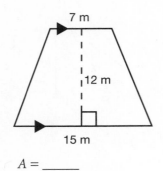

7 m

12 m

15 m

$A =$ _____

25.

b_1

9 cm

15 cm

$A = 90$ cm^2

$b_1 =$ _____

26.

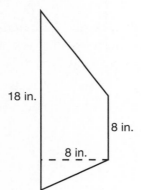

18 in.

8 in.

8 in.

27.

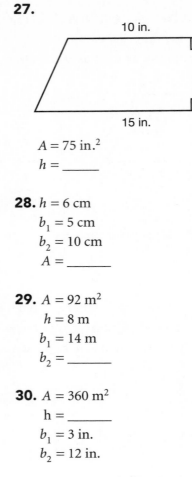

10 in.

h

15 in.

$A = 75$ in.^{2}

$h =$ _____

28. $h = 6$ cm

$b_1 = 5$ cm

$b_2 = 10$ cm

$A =$ _____

29. $A = 92$ m^2

$h = 8$ m

$b_1 = 14$ m

$b_2 =$ _____

30. $A = 360$ m^2

h = _____

$b_1 = 3$ in.

$b_2 = 12$ in.

Skill Building until Next Time

Try to find an object shaped like each of the polygons in this lesson. Examples would be a rectangular window pane, a square floor tile, a table leg shaped like a trapezoid, a quilt piece shaped like a parallelogram, and a triangular scarf. Measure and find the area of each.

14 ▶ SURFACE AREA OF PRISMS

Lesson Summary

In this lesson, you will learn how to identify various parts of a prism and how to use formulas to find the surface area of a prism.

A rectangular prism is a shape you see every day. Examples are bricks, cereal boxes, and some buildings. Sometimes, it is necessary to completely cover these objects. You might need to wrap a gift, determine how much siding is needed to cover a house, or make a label for a box.

A rectangular prism has six faces. The faces are parts of planes that form sides of solid figures. A rectangular prism has 12 edges. The edges are the segments formed by the intersection of two faces of a solid figure. A rectangular prism has eight vertices. Vertices are the points where the edges meet.

Example:

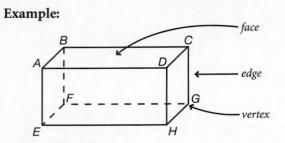

ABCD is a face.

CG is an edge.

G is a vertex.

Edges that are parallel to each other have the same measurements.

Example:

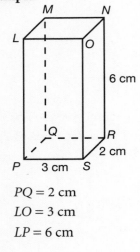

PQ = 2 cm

LO = 3 cm

LP = 6 cm

Practice

Determine the length of each side.

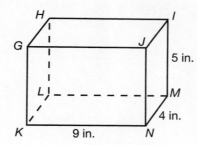

1. $HL =$ _____

2. $KL =$ _____

3. $GJ =$ _____

4. $GK =$ _____

5. $JI =$ _____

6. $LM =$ _____

7. How many edges does a rectangular solid have?

8. How many faces does a rectangular solid have?

9. How many vertices does a rectangular solid have?

10. How many vertices does any face touch?

11. How many vertices have a length of 5 in.?

12. How many edges come together at each vertex?

Using a Formula to Find the Surface Area of a Rectangular Prism

If you take a box apart and fold out all the sides, you will have six rectangles. You will have three pairs that are the same size. You could find the area of each rectangle and add up all their areas to obtain the total area.

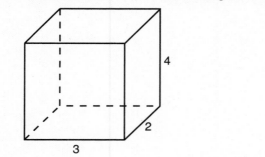

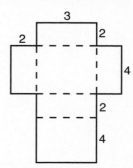

You could make a chart to help you organize all the faces, or you could use a formula.

Theorem: The surface area (*S.A.*) of a rectangular prism is twice the sum of the length (*l*) times the width (*w*), the width (*w*) times the height (*h*), and the length (*l*) times the height (*h*).

$$S.A. = 2(lw + wh + lh)$$

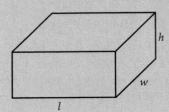

Example: Find the surface area of the rectangular prism.

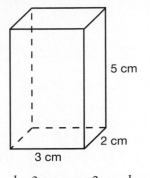

$l = 3$ cm, $w = 2$ cm, $h = 5$ cm

$S.A. = 2(lw + wh + lh)$

$S.A. = 2[(3)(2) + (2)(5) + (3)(5)]$

$S.A. = 2(6 + 10 + 15)$

$S.A. = 2(31)$

$S.A. = 62$ cm^2

Practice

Find the surface area of each rectangular prism.

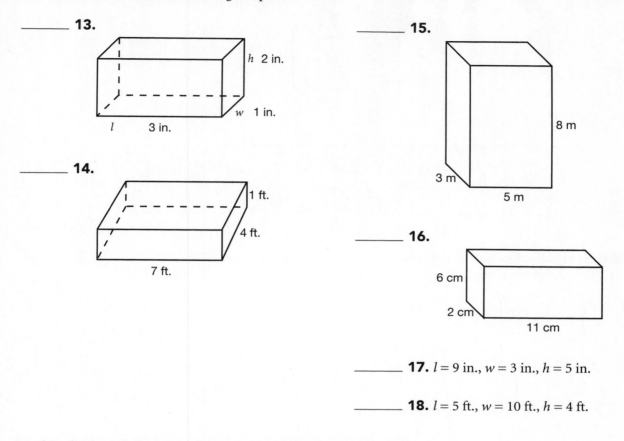

_____ **13.**

h 2 in.
w 1 in.
l 3 in.

_____ **14.**

1 ft.
4 ft.
7 ft.

_____ **15.**

8 m
3 m
5 m

_____ **16.**

6 cm
2 cm
11 cm

_____ **17.** $l = 9$ in., $w = 3$ in., $h = 5$ in.

_____ **18.** $l = 5$ ft., $w = 10$ ft., $h = 4$ ft.

Working backward, find the missing edge length using the given information.

19. $A = 448$ ft.^{2}
$l = 14$ ft.
$w = ?$
$h = 6$ ft.

20. $A = 280$ m^2
$l = 4$ m
$w = 4$ m
$h = ?$

The Surface Area of a Cube

A cube is a special rectangular prism. All the edges of a cube have the same length, so all six faces have the same area.

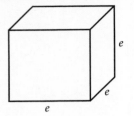

> *Theorem:* The surface area of a cube is six times the edge (e) squared.
>
> $$S.A. = 6e^2$$

Example: Find the surface area of the cube.

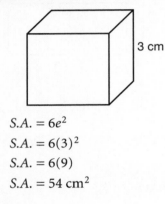

3 cm

$S.A. = 6e^2$

$S.A. = 6(3)^2$

$S.A. = 6(9)$

$S.A. = 54 \text{ cm}^2$

Practice

Find the surface area of each cube.

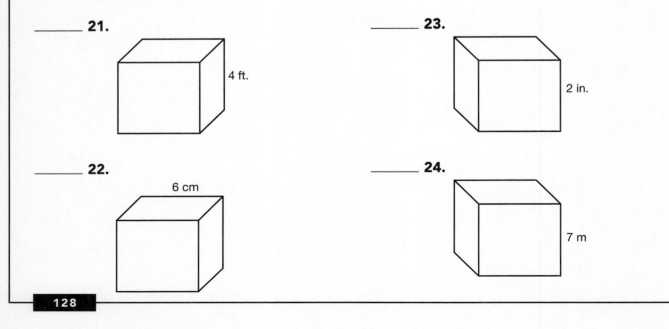

_____ **21.**

4 ft.

_____ **23.**

2 in.

_____ **22.**

6 cm

_____ **24.**

7 m

_____ **25.** $e = 5.8$ cm

_____ **26.** If the surface area of a cube is 486 ft.^{2}, find the length of each edge.

Working backward, find the missing edge length using the given information.

_____ **27.** If the surface area of a cube is 726 in.^{2}, find the length of one of its edges.

_____ **28.** If the surface area of a cube is 73.5 ft.^{2}, find the perimeter of its front face.

Use the following information to answer questions 29 and 30.

Susie bought a blender that came in a rectangular prism box with edges measuring 8 in., 12 in., and 18 in. She needs to wrap it as a gift and wants to use her blue wrapping paper, which measures 9 in. × 100 in. If she doesn't have enough blue, she will use her green paper, which is 12 in. × 80 in. If she doesn't have enough green, she will use her yellow paper, of which she has an endless supply. (Assume that there's no need for the edges of the wrapping paper to overlap; Susie simply needs to cover the entire box.)

_____ **29.** How many square inches of wrapping paper will Susie need to cover the blender box?

_____ **30.** What color paper will Susie use?

Skill Building until Next Time

Take a juice box and measure its length, width, and height. Use the formula to find its surface area. Now take a break and drink the juice. Unfold the empty juice box. Find the area of each rectangle that made up the box. Compare the two answers.

15 ▶ VOLUME OF PRISMS AND PYRAMIDS

Lesson Summary

In this lesson, you will learn how to find the volume of prisms and pyramids using formulas.

 hen you are interested in finding out how much a refrigerator holds or the amount of storage space in a closet, you are looking for the volume of a prism. Volume is expressed in cubic units. Just like an ice tray is filled with cubes of ice, volume tells you how many cubic units will fit into a space.

Volume of Prisms

Prisms can have bases in the shape of any polygon. A prism with a rectangle for its bases is called a rectangular prism. A prism with triangles for its bases is referred to as a triangular prism. A prism with hexagons as its bases is called a hexagonal prism, and so on. Prisms can be right or oblique. A right prism is a prism with its bases perpendicular to its sides, meaning they form right angles. Bases of oblique prisms do not form right angles with their sides. An example of an oblique prism is the Leaning Tower of Pisa. In this lesson, you will concentrate on the volume of right prisms. When we refer to prisms in this lesson, you can assume we mean a right prism.

> *Theorem:* To find the volume (V) of a rectangular prism, multiply the length (l) by the width (w) and by the height (h).
> $$V = lwh$$

The volume of a rectangular prism could also be stated in another way. The area of the base of a rectangular prism is the length times the width, the same length and width used with the height to find the volume—in other words, the area of the base (B) times the height. This same approach can also be applied to other solid figures. In Lesson 13, you studied ways to find the area of most polygons, but not all. However, if you are given the area of the base and the height of the figure, then it is possible to find its volume. Likewise, if you do know how to find the area of the base (B), use the formula for that shape, then multiply by the height (h) of the figure.

> *Theorem:* To find the volume (V) of any prism, multiply the area of a base (B) by the height (h).
> $$V = Bh$$

Examples: Find the volume of each prism.

(a)

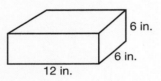

12 in. 6 in. 6 in.

$V = lwh$
$V = (12)(6)(6)$
$V = 432$ in.^{3}

(b)

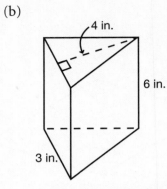

Solution: First, find the area of a base. As you remember from Lesson 13, the formula for finding the area of any triangle is $A = \frac{1}{2}bh$.

$$A = \frac{1}{2}(3)(4)$$
$$A = \frac{1}{2}(12)$$
$$A = 6 \text{ in.}^2$$

The area, A, becomes the base, B, in the volume formula $V = Bh$.

$$V = Bh$$
$$V = (6)(6)$$
$$V = 36 \text{ in.}^3$$

(c)

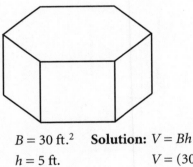

$B = 30 \text{ ft.}^2$ **Solution:** $V = Bh$

$h = 5 \text{ ft.}$ $V = (30)(5)$

                 $V = 150 \text{ ft.}^3$

(d)   Prism: $B = 42 \text{ m}^2$, $h = 10 \text{ m}$ **Solution:** $V = Bh$

                                    $V = (42)(10)$

                                    $V = 420 \text{ m}^3$

Practice

Find the volume of each prism.

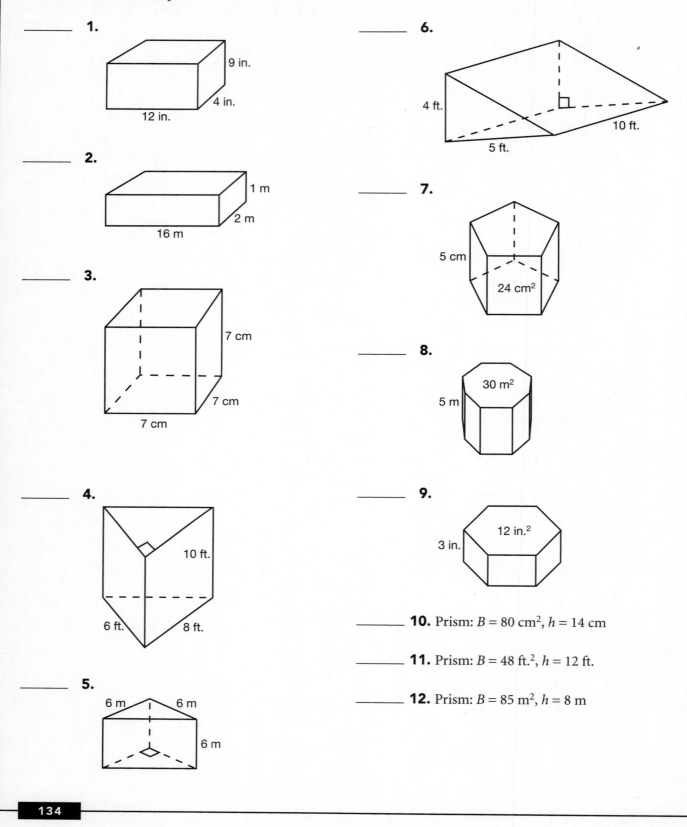

_____ 1.

9 in.
4 in.
12 in.

_____ 2.

1 m
2 m
16 m

_____ 3.

7 cm
7 cm
7 cm

_____ 4.

10 ft.
6 ft.
8 ft.

_____ 5.

6 m
6 m
6 m

_____ 6.

4 ft.
5 ft.
10 ft.

_____ 7.

5 cm
24 cm²

_____ 8.

30 m²
5 m

_____ 9.

12 in.²
3 in.

_____ 10. Prism: $B = 80$ cm², $h = 14$ cm

_____ 11. Prism: $B = 48$ ft.², $h = 12$ ft.

_____ 12. Prism: $B = 85$ m², $h = 8$ m

For questions 13 and 14, use the given volume to determine the length of the missing edge.

13. In the following figure, the area of the rectangular base is 52 cm^2 and the total volume is 468 cm^3.

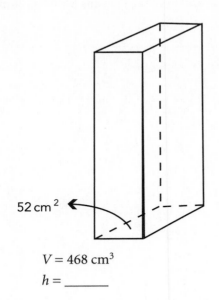

52 cm^2

$V = 468$ cm^3

$h = $ _____

14. The following figure has a height of 6 in., a length of 8 in., and a total volume of 336 in.^{3}.

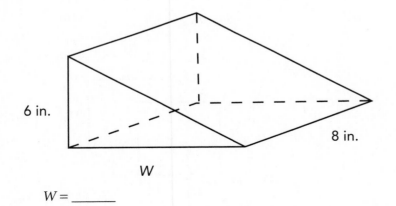

6 in.

8 in.

W

$W = $ _____

Volume of a Cube

Recall that the edges (e) of a cube all have the same measurement; therefore, if you replace the length (l), width (w), and height (h) with the measurement of the edge of the cube, then you will have the formula for the volume of the cube, $V = e^3$.

> *Theorem:* The volume of a cube is determined by cubing the length of the edge.
> $$V = e^3$$

Example: Find the volume of the cube.

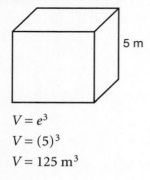

5 m

$$V = e^3$$
$$V = (5)^3$$
$$V = 125 \text{ m}^3$$

Practice

Find the volume of each cube.

_____ **15.**

_____ **16.**

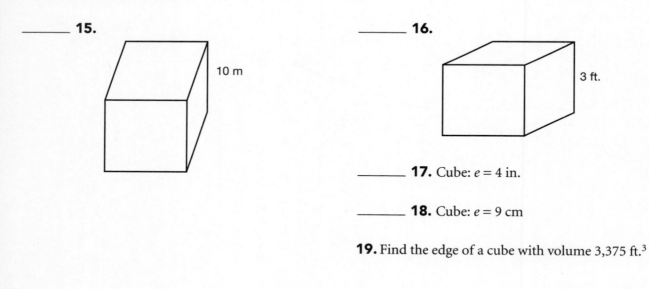

10 m

3 ft.

_____ **17.** Cube: $e = 4$ in.

_____ **18.** Cube: $e = 9$ cm

19. Find the edge of a cube with volume 3,375 ft.³

Volume of a Pyramid

A polyhedron is a three-dimensional figure whose surfaces are all polygons. A regular pyramid is a polyhedron with a base that is a regular polygon and a vertex point that lies directly over the center of the base.

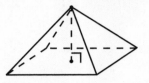

If you have a pyramid and a rectangular prism with the same length, width, and height, you would find that it would take three of the pyramids to fill the prism. In other words, one-third of the volume of the prism is the volume of the pyramid.

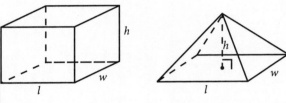

$$V = lwh \text{ or } V = Bh \qquad\qquad V = \tfrac{1}{3}lwh \text{ or } V = \tfrac{1}{3}Bh$$

Examples: Find the volume of each pyramid.

(a)

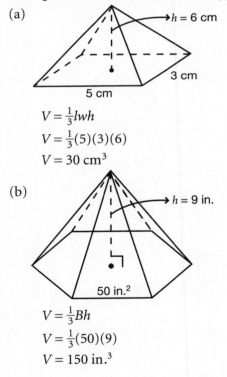

$h = 6$ cm

3 cm

5 cm

$V = \tfrac{1}{3}lwh$

$V = \tfrac{1}{3}(5)(3)(6)$

$V = 30$ cm^3

(b)

$h = 9$ in.

50 in.^{2}

$V = \tfrac{1}{3}Bh$

$V = \tfrac{1}{3}(50)(9)$

$V = 150$ in.^{3}

(c) Regular pyramid: $B = 36$ ft.^{2}, $h = 8$ ft.   **Solution:** $V = \tfrac{1}{3}Bh$

$V = \tfrac{1}{3}(36)(8)$

$V = 96$ ft.^{3}

Practice

Find the volume of each pyramid.

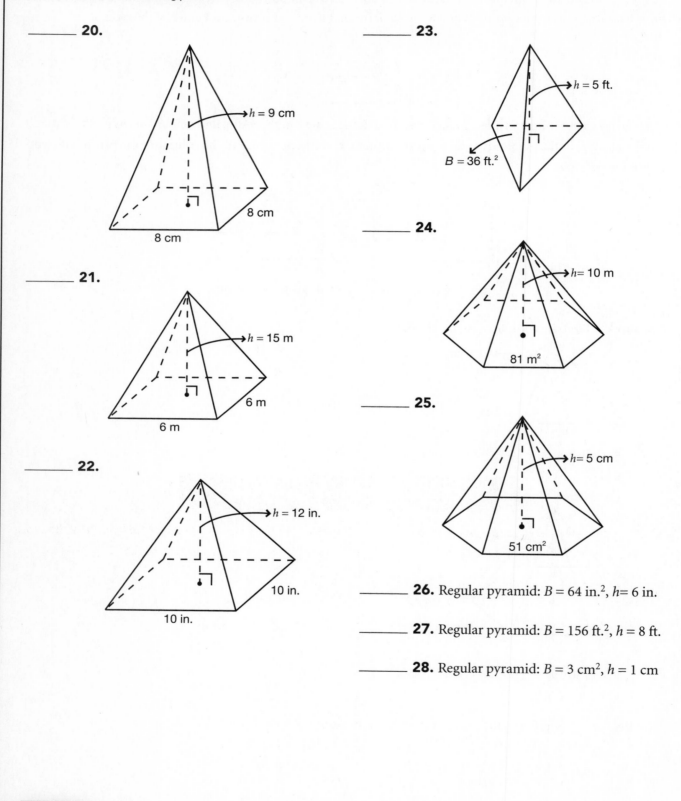

_____ **20.**

$h = 9$ cm

8 cm

8 cm

_____ **21.**

$h = 15$ m

6 m

6 m

_____ **22.**

$h = 12$ in.

10 in.

10 in.

_____ **23.**

$h = 5$ ft.

$B = 36$ ft.²

_____ **24.**

$h = 10$ m

81 m²

_____ **25.**

$h = 5$ cm

51 cm²

_____ **26.** Regular pyramid: $B = 64$ in.², $h = 6$ in.

_____ **27.** Regular pyramid: $B = 156$ ft.², $h = 8$ ft.

_____ **28.** Regular pyramid: $B = 3$ cm², $h = 1$ cm

For questions 29 and 30, use the given volume to determine the missing value.

29.

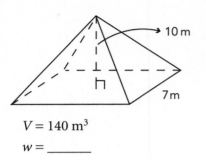

$V = 140 \text{ m}^3$

$w = \underline{\hspace{1cm}}$

30.

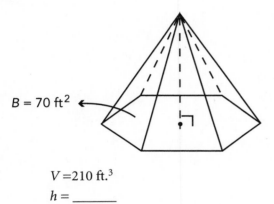

$B = 70 \text{ ft}^2$

$V = 210 \text{ ft.}^3$

$h = \underline{\hspace{1cm}}$

Skill Building until Next Time

Take a box of saltine crackers, the kind with four packages of crackers inside. Measure the length, width, and height of the cracker box and find its volume. Now measure the length, width, and height of the sleeve of crackers and find its volume. Compare the two volumes.

16 ▶ WORKING WITH CIRCLES AND CIRCULAR FIGURES

Lesson Summary

In this lesson, you will learn about the irrational number pi, or π. You will also learn to use formulas to find the circumference and area of circles, the surface area and volume of cylinders and spheres, and the volume of cones.

Before you begin to work with circles and circular figures, you need to know about the irrational number, π (pronounced "pie"). Over 2,000 years ago, mathematicians approximated the value of the ratio of the distance around a circle to the distance across a circle to be approximately 3. Years later, this value was named with the Greek letter π. The exact value of π is still a mathematical mystery. π is an irrational number. A *rational number* is a number that can be written as a ratio, a fraction, or a terminating or repeating decimal. Although its value has been computed in various ways over the past several hundred years, no one has been able to find a decimal value of π where the decimal terminates or develops a repeating pattern. Computers have been used to calculate the value of π to over fifty billion decimal places, but there is still no termination or repeating group of digits.

$$\pi = \frac{circumference}{diameter}$$

The most commonly used approximations for π are $\frac{22}{7}$ and 3.14. These are not the true values of π, only rounded approximations. You may have a π key on your calculator. This key will give you an approximation for π that varies according to how many digits your calculator displays.

Circumference of a Circle

Now that you know about the irrational number π, it's time to start working with circles. The distance around a circle is called its *circumference.* There are many reasons why people need to find a circle's circumference. For example, the amount of lace edge around a circular skirt can be found by using the circumference formula. The amount of fencing for a circular garden is another example of when you need the circumference formula.

Since π is the ratio of circumference to diameter, the approximation of π times the diameter of the circle gives you the circumference of the circle. The diameter of a circle is the distance across a circle through its center. A radius is the distance from the center to the edge. One-half the diameter is equal to the radius or two radii are equal to the length of the diameter.

$$d = 2r \text{ or } r = \tfrac{1}{2}d$$

Here is a theorem that will help you solve circumference problems:

> *Theorem:* The circumference of any circle is the product of its diameter and π.
> $$C = \pi d \text{ or } C = 2\pi r$$

Since π is approximately (not exactly) equal to 3.14, after you substitute the value 3.14 for π in the formula, you should use ≈ instead of =. The symbol ≈ means *approximately equal to.*

Examples: Find the circumference of each circle. Use the approximation of 3.14 for π.

(a)

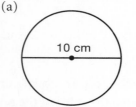

10 cm

Solution: Use the $C = \pi d$ formula, since the diameter of the circle is given.

$C = \pi d$

$C \approx (3.14)(10)$

$C \approx 31.4$ cm

(b)

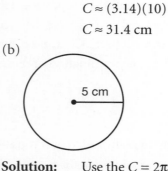

5 cm

Solution: Use the $C = 2\pi r$ formula, since the radius of the circle is given.

$C = 2\pi r$

$C \approx 2(3.14)(5)$

$C \approx 31.4$ cm

Notice that these two circles have the same circumference because a circle with a diameter of 10 cm has a radius of 5 cm. You pick which formula to use based on what information you are given—the circle's radius or its diameter.

Practice

Find the approximate circumference of each circle shown or described. Use 3.14 for π.

_____ **1.**

_____ **3.** $d = 7$ m

_____ **4.** $r = 25$ m

_____ **5.**

_____ **2.**

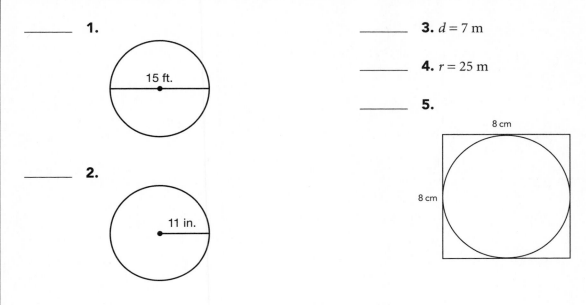

Area of a Circle

To understand the area of a circle, take a look at the following figure. Imagine a circle that is cut into wedges and rearranged to form a shape that resembles a parallelogram.

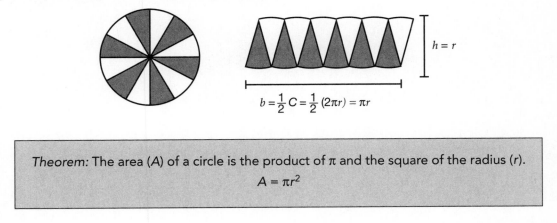

> *Theorem:* The area (A) of a circle is the product of π and the square of the radius (r).
> $$A = \pi r^2$$

Notice that this formula squares the radius, not the diameter, so if you are given the diameter, you should divide it by two or multiply it by one-half to obtain the radius. Some people will mistakenly think that squaring the radius and doubling it to get the diameter are the same. They are the same only when the radius is 2. Otherwise, two times a number and the square of a number are very different.

Examples: Find the approximate area for each circle. Use 3.14 for π.

(a)

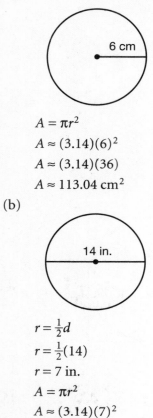

$A = \pi r^2$

$A \approx (3.14)(6)^2$

$A \approx (3.14)(36)$

$A \approx 113.04 \text{ cm}^2$

(b)

$r = \frac{1}{2}d$

$r = \frac{1}{2}(14)$

$r = 7 \text{ in.}$

$A = \pi r^2$

$A \approx (3.14)(7)^2$

$A \approx (3.14)(49)$

$A \approx 153.86 \text{ in.}^2$

Practice
Find the approximate area for each circle shown or described. Use 3.14 for π.

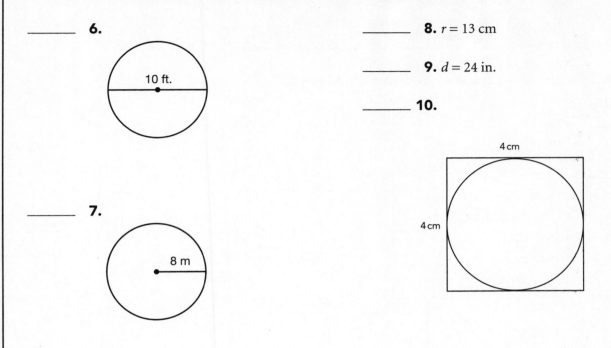

_____ **6.**

10 ft.

_____ **7.**

8 m

_____ **8.** $r = 13$ cm

_____ **9.** $d = 24$ in.

_____ **10.**

4 cm

4 cm

Surface Area of a Cylinder

When you are looking for the surface area of a cylinder, you need to find the area of two circles (the bases) and the area of the curved surface that makes up the side of the cylinder. The area of the curved surface is hard to visualize when it is rolled up. Picture a paper towel roll. It has a circular top and bottom. When you unroll a sheet of the paper towel, it is shaped like a rectangle. The area of the curved surface is the area of a rectangle with the same height as the cylinder, and the base measurement is the same as the circumference of the circle base.

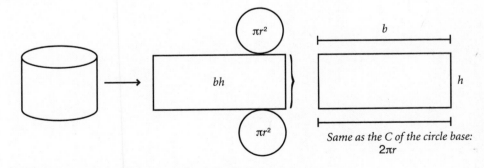

Same as the C of the circle base:
$2\pi r$

Surface area of a cylinder = area of two circles + area of rectangle

$$= 2\pi r^2 + bh$$
$$= 2\pi r^2 + 2\pi rh$$

> *Theorem:* The surface area (*S.A.*) of a cylinder is determined by finding the sum of the area of the bases and the product of the circumference times the height.
>
> $$S.A. = 2\pi r^2 + 2\pi rh$$

Examples: Find the surface area of each cylinder. Use 3.14 for π.

(a)

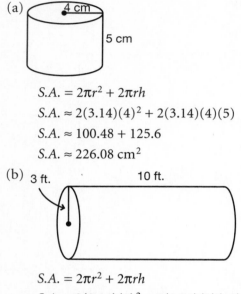

$S.A. = 2\pi r^2 + 2\pi rh$

$S.A. \approx 2(3.14)(4)^2 + 2(3.14)(4)(5)$

$S.A. \approx 100.48 + 125.6$

$S.A. \approx 226.08 \text{ cm}^2$

(b)

$S.A. = 2\pi r^2 + 2\pi rh$

$S.A. \approx 2(3.14)(3)^2 + 2(3.14)(3)(10)$

$S.A. \approx 56.52 + 188.4$

$S.A. \approx 244.92 \text{ ft.}^2$

Practice

Find the surface area of each cylinder. Use 3.14 for π.

_____ **11.**

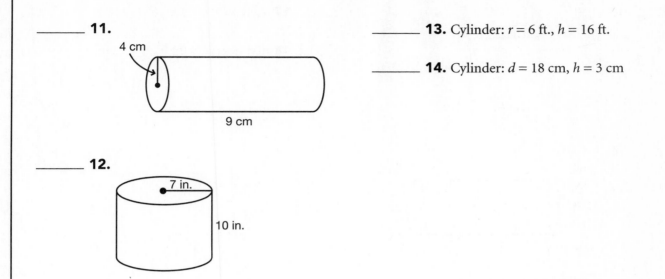

_____ **13.** Cylinder: $r = 6$ ft., $h = 16$ ft.

_____ **14.** Cylinder: $d = 18$ cm, $h = 3$ cm

_____ **12.**

Volume of a Cylinder

Similar to finding the volume of a prism, you can find the volume of a cylinder by finding the product of the area of the base and the height of the figure. Of course, the base of a cylinder is a circle, so you need to find the area of a circle times the height.

> _Theorem:_ The volume (V) of a cylinder is the product of the area of the base (B) and the height (h).
> $$V = Bh \text{ or } V = \pi r^2 h$$

Examples: Find the volume of each cylinder. Use 3.14 for π.

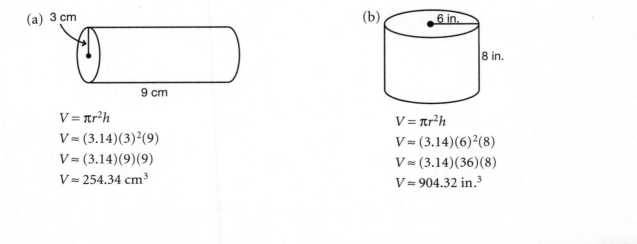

(a)

$V = \pi r^2 h$
$V \approx (3.14)(3)^2(9)$
$V \approx (3.14)(9)(9)$
$V \approx 254.34 \text{ cm}^3$

(b)

$V = \pi r^2 h$
$V \approx (3.14)(6)^2(8)$
$V \approx (3.14)(36)(8)$
$V \approx 904.32 \text{ in.}^3$

Practice

Find the volume of each cylinder shown or described. Use 3.14 for π.

_____ **15.**

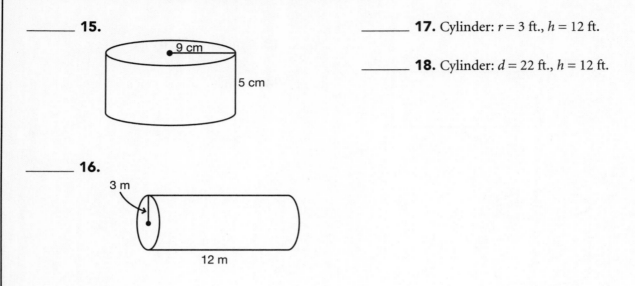

_____ **17.** Cylinder: $r = 3$ ft., $h = 12$ ft.

_____ **18.** Cylinder: $d = 22$ ft., $h = 12$ ft.

_____ **16.**

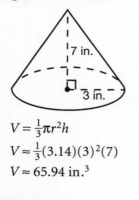

Volume of a Cone

A cone relates to a cylinder in the same way that a pyramid relates to a prism. If you have a cone and a cylinder with the same radius and height, it would take three of the cones to fill the cylinder. In other words, the cone holds one-third the amount of the cylinder.

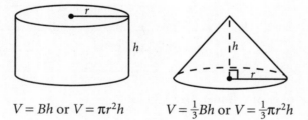

$$V = Bh \text{ or } V = \pi r^2 h \qquad V = \tfrac{1}{3}Bh \text{ or } V = \tfrac{1}{3}\pi r^2 h$$

Example: Find the volume of the cone. Use 3.14 for π.

$$V = \tfrac{1}{3}\pi r^2 h$$
$$V \approx \tfrac{1}{3}(3.14)(3)^2(7)$$
$$V \approx 65.94 \text{ in.}^3$$

Practice

Find the volume of each cone shown or described. Use 3.14 for π.

_____ **19.**

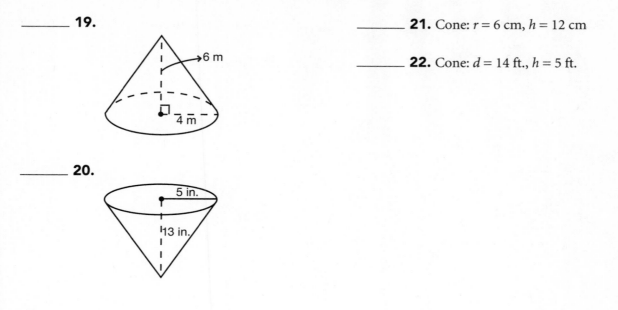

_____ **21.** Cone: $r = 6$ cm, $h = 12$ cm

_____ **22.** Cone: $d = 14$ ft., $h = 5$ ft.

_____ **20.**

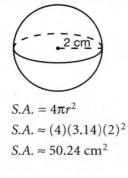

Surface Area of a Sphere

A *sphere* is the set of all points that are the same distance from some point called the center. A sphere is most likely to be called a ball. Try to find an old baseball and take the cover off of it. When you lay out the cover of the ball, it roughly appears to be four circles. Recall that the formula for finding the area of a circle is $A = \pi r^2$.

> *Theorem:* The surface area (S.A.) formula for a sphere is four times π times the radius squared.
> $$S.A. = 4\pi r^2$$

Example: Find the surface area of the sphere. Use 3.14 for π.

$S.A. = 4\pi r^2$

$S.A. \approx (4)(3.14)(2)^2$

$S.A. \approx 50.24 \text{ cm}^2$

Practice

Find the surface area of each sphere shown or described. Use 3.14 for π.

_____ **23.**

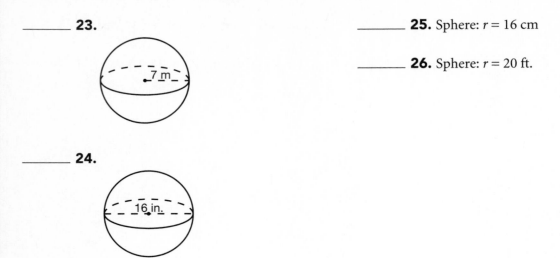

7 m

_____ **25.** Sphere: $r = 16$ cm

_____ **26.** Sphere: $r = 20$ ft.

_____ **24.**

16 in.

Volume of a Sphere

If you were filling balloons with helium, it would be important for you to know the volume of a sphere. To find the volume of a sphere, picture the sphere filled with numerous pyramids. The height of each pyramid represents the radius (r) of the sphere. The sum of the areas of all the bases represents the surface area of the sphere.

Volume of each pyramid = $\frac{1}{3}Bh$

Sum of the volumes = $n \times \frac{1}{3}Br$ Substitute r for h of n pyramids

$\qquad\qquad\qquad = \frac{1}{3}(nB)r$ Substitute nB with the $S.A.$ of a sphere

$\qquad\qquad\qquad = \frac{1}{3}(4\pi r^2)r$

$\qquad\qquad\qquad = \frac{4}{3}\pi r^3$

> _Theorem:_ The volume (V) of a sphere is determined by the product of $\frac{4}{3}\pi$ and the cube of the radius.
>
> $$V = \frac{4}{3}\pi r^3$$

Example: Find the volume of the sphere. Use 3.14 for π.

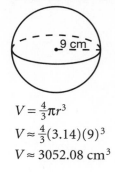

$V = \frac{4}{3}\pi r^3$

$V \approx \frac{4}{3}(3.14)(9)^3$

$V \approx 3052.08 \text{ cm}^3$

Practice

Find the volume of each sphere. Use 3.14 for π.

_____ **27.**

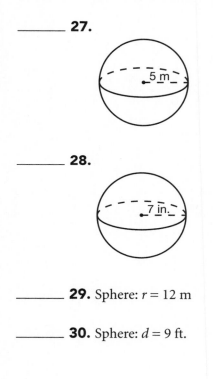

_____ **28.**

_____ **29.** Sphere: $r = 12$ m

_____ **30.** Sphere: $d = 9$ ft.

Skill Building until Next Time

Find an example of a cylinder—an oatmeal box would be a good choice. Measure the height and radius of the cylinder. Use the formulas in this lesson to find the surface area and volume of the cylinder.

17 ▶ COORDINATE GEOMETRY

LESSON SUMMARY

In this lesson, you will learn to identify the *x*-axis, *y*-axis, the origin, and the four quadrants on a coordinate plane. You will also learn how to plot or graph points on a coordinate plane and name the coordinates of a point. The distance between two points will also be found using a formula.

I f you have ever been the navigator on a road trip, then you have probably read a road map or grid map. A grid map uses a horizontal and vertical axis in a similar manner as a coordinate plane.

On a coordinate plane, the horizontal axis is called the *x*-axis. The vertical axis is called the *y*-axis. The point where the two axes cross is called the *origin*.

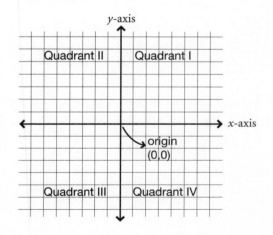

The two axes divide the coordinate plane into four regions, which are called *quadrants*. The quadrants are numbered counterclockwise beginning with the upper-right region. The coordinates (x,y) of a point are an ordered pair of numbers. The first number is the x-coordinate. The second number is the y-coordinate. The coordinates of the origin are $(0,0)$.

Example: Graph point $A(-4,1)$ and point $B(5,-3)$. In which quadrant would you find each point?

Solution:

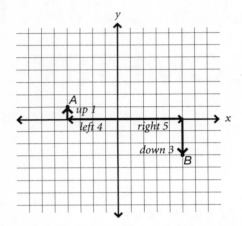

To graph point $A(-4,1)$, from the origin, go left 4 units and up 1. Label the point A. Point A is in quadrant II. To graph point $B(5,-3)$, start from the origin, then go right 5 units and down 3. Label the point B. Point B is in quadrant IV.

Some points may be graphed on the axes also.

Example: Graph point $C(2,0)$ and point $D(0,-6)$. On which axis will each point lie?

Solution:

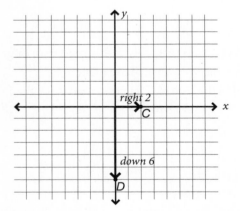

To graph point C, from the origin, go right 2 units, but do not move up or down. Label the point C. Point C is on the x-axis. To graph point D, from the origin, do not move right or left; move 6 units down. Label the point D. Point D is on the y-axis.

Practice

Draw a set of axes on graph paper. Graph and label each point.

1. $A(-6,2)$ **7.** $G(1,4)$

2. $B(5,3)$ **8.** $H(-2,5)$

3. $C(4,-5)$ **9.** $I(-1,3)$

4. $D(-1,-1)$ **10.** $J(5,-3)$

5. $E(0,6)$ **11.** $K(2,0)$

6. $F(-3,0)$ **12.** $L(0,-5)$

In which quadrant or axis does each of the following points lie?

_____ **13.** $A(-6,2)$ _____ **19.** $G(1,4)$

_____ **14.** $B(6,4)$ _____ **20.** $H(0,0)$

_____ **15.** $C(4,-5)$ _____ **21.** $I(-1,-1)$

_____ **16.** $D(-5,-5)$ _____ **22.** $J(3,-2)$

_____ **17.** $E(0,6)$ _____ **23.** $K(2,0)$

_____ **18.** $F(-3,0)$ _____ **24.** $L(0,-5)$

Finding the Coordinates of a Point

Each point on the coordinate plane has its own unique ordered pair. You can think of an ordered pair as an address. Now that you have located a point, you can also find the coordinates of a point on a graph.

Example: Find the coordinates of each point.

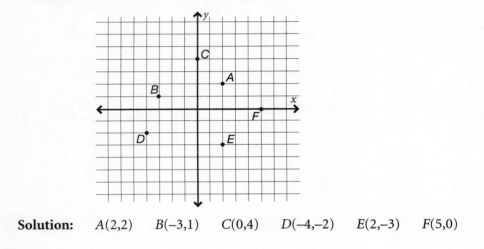

Solution: $A(2,2)$ $B(-3,1)$ $C(0,4)$ $D(-4,-2)$ $E(2,-3)$ $F(5,0)$

Practice

Find the coordinates of each point.

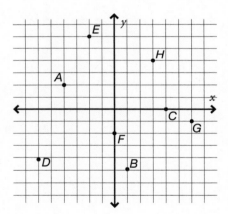

_____ **25.** A

_____ **26.** B

_____ **27.** C

_____ **28.** D

_____ **29.** E

_____ **30.** F

_____ **31.** G

_____ **32.** H

Finding the Distance between Two Points

You can easily count to find the distance between two points on a horizontal or vertical line. For example, in the following figure, $\overline{XY} = 3$ and $\overline{YZ} = 4$. However, you cannot find $\overline{XZ}$ simply by counting. Since $\triangle XYZ$ is a right triangle, you can use the Pythagorean theorem.

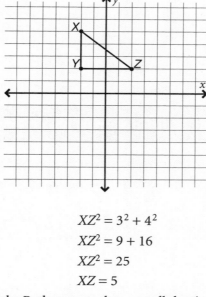

$$XZ^2 = 3^2 + 4^2$$
$$XZ^2 = 9 + 16$$
$$XZ^2 = 25$$
$$XZ = 5$$

Of course, you won't be able to use the Pythagorean theorem all the time. However, you can use the following formula to find the distance between any two points.

> **Theorem:** The distance d between any two points $A(x_1, y_1)$ and $B(x_2, y_2)$ is
> $$d = \sqrt{(x_2 - x_1)^2 + (y_2 - y_1)^2}$$

Example: Find the distance between $R(-3,4)$ and $S(-3,-7)$.

Solution: It helps to label the coordinates of the points before you insert them into the formula.

$$\begin{array}{cc} x_1 y_1 & x_2 y_2 \\ R(-3,4) & S(-3,-7) \end{array}$$

$$d = \sqrt{(x_2 - x_1)^2 + (y_2 - y_1)^2}$$
$$d = \sqrt{(-3 - (-3))^2 + (-7 - 4)^2}$$
$$d = \sqrt{(-3 + 3)^2 + (-11)^2}$$
$$d = \sqrt{0^2 + 121}$$
$$d = \sqrt{121}$$
$$d = 11$$

Practice

Find the distance between the points with the given coordinates.

_____ **33.** (3,5) and (–1,5)

_____ **34.** (3,–2) and (–5,4)

_____ **35.** (–3,–8) and (–6,–4)

_____ **36.** (–1,–6) and (4,6)

Skill Building until Next Time

Examine a city map to see if there are any streets that cut diagonally across horizontal and vertical streets. If so, construct a coordinate plane for the map and determine the distance between two locations that are on the diagonal street.

18 ▶ THE SLOPE OF A LINE

LESSON SUMMARY

In this lesson, you will learn how to determine the slope of a line from its graph or from two points on the line. You will also learn how to tell by sight if the slope of a line is positive, negative, zero, or undefined.

The *slope* of a line is the measure of its steepness. The slope of a line is determined by the ratio of its rise to run. When looking at a line on a coordinate grid, always count the run before you count the rise. When a line points up to the right, it has a positive slope. A line with a negative slope points up to the left.

Examples: Find the slope of each line.

(a)

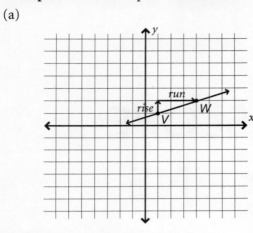

Solution: Slope of $\overline{VW} = \frac{rise}{run} = \frac{1}{3}$

(b)

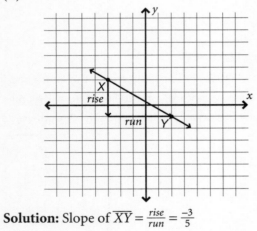

Solution: Slope of $\overline{XY} = \frac{rise}{run} = \frac{-3}{5}$

Practice

Find the slope of each line.

_____ **1.**

_____ **4.**

_____ **2.**

_____ **5.**

_____ **3.**

_____ **6.**

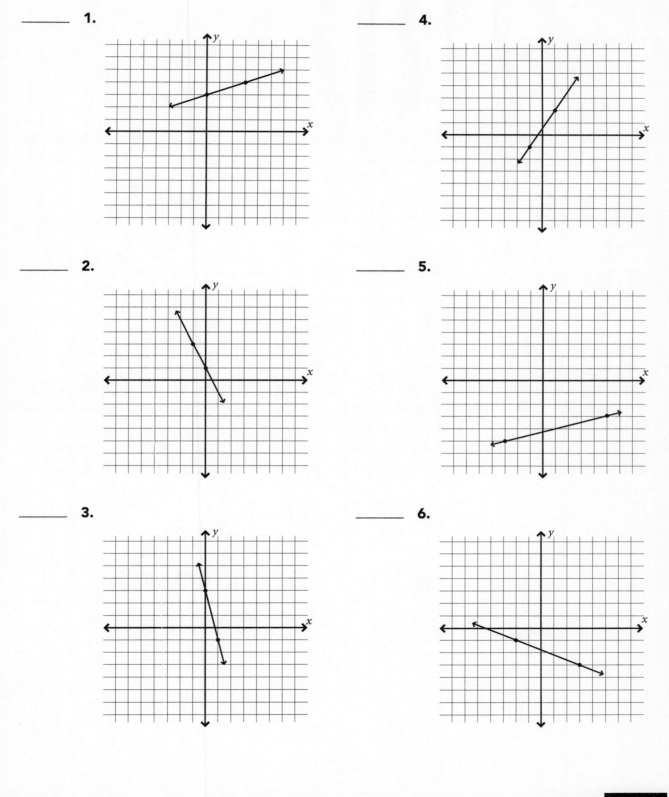

For questions 7 and 8, graph the line made by the given coordinates and find its slope.

7. (3,2) and (−7,6)

8. (−1,5) and (3,−5)

Special Cases of Slope

You may be wondering what the slope is for horizontal and vertical lines—these two types of lines do not slant up toward the left or the right. They are special cases.

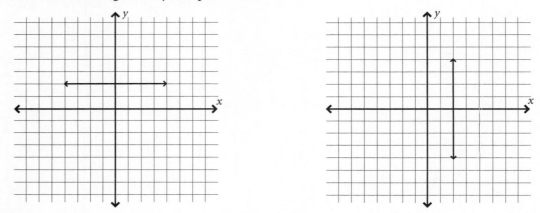

The horizontal line has a slope of zero, since the rise is equal to zero. (Remember that zero divided by anything equals zero, so $\frac{rise}{run} = \frac{0}{x} = 0$.) On the other hand, the vertical line has an undefined slope (sometimes said to have "no slope"). In this case, the run is equal to zero, so $\frac{rise}{run} = \frac{y}{0}$, and a number divided by zero is said to be *undefined*, since it doesn't make sense to take something and divide it up into zero parts. The following illustrations may help you remember which type of line—horizontal or vertical—has an undefined (no slope) or a slope of zero.

Practice

Tell whether the slope of each line is positive, negative, zero, or undefined (no slope.

9.

10.

11.

12.

13.

14.

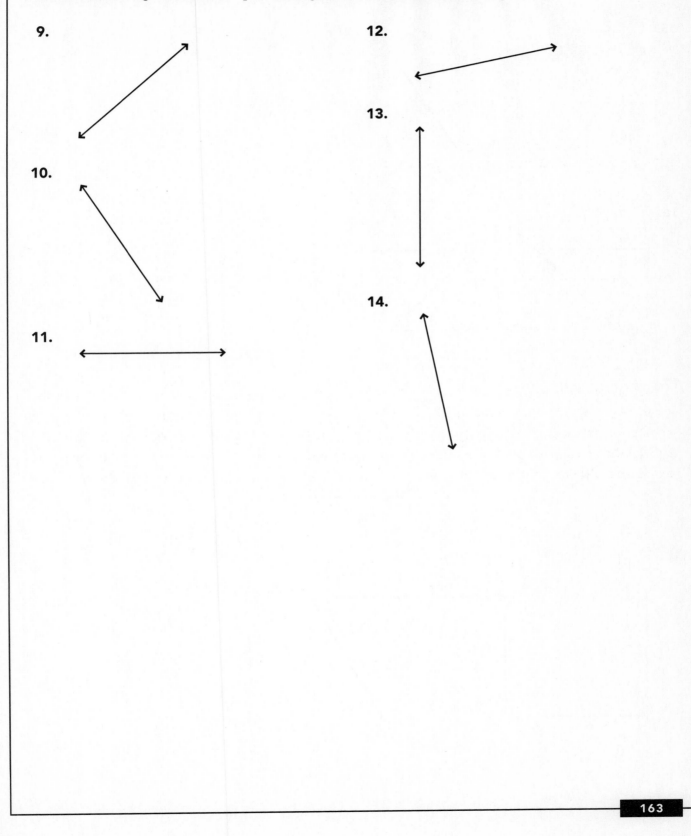

For questions 15–18 use the shapes in the following figures and label the type of slope of each line segment, moving from left to right.

15.

16.

17.

18.

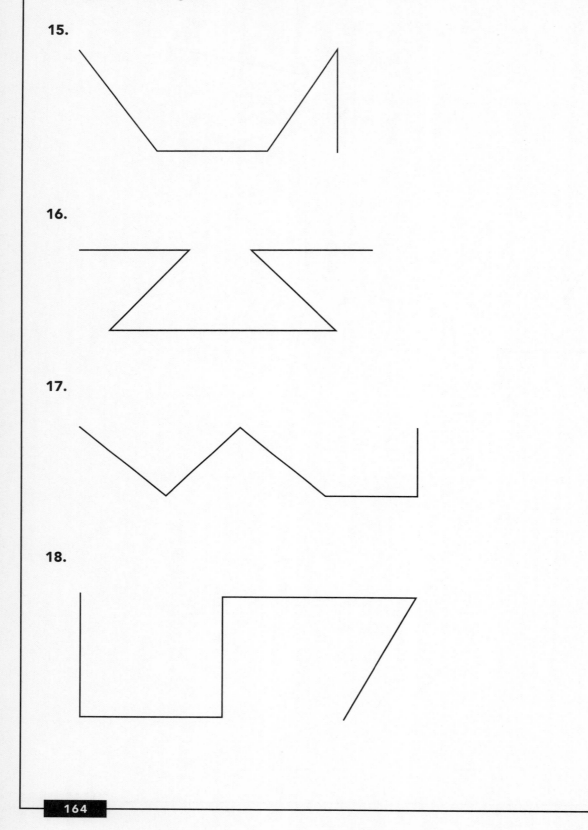

Using a Formula

You can also use a formula to determine the slope of a line containing two points, point $A(x_1, y_1)$ and point $B(x_2, y_2)$. Here is the formula:

$$\text{slope} = \frac{y_2 - y_1}{x_2 - x_1}$$

Example: Find the slope of the line through $P(-6,5)$ and $Q(-2,-1)$.

Solution: Begin by labeling the coordinates before you insert the values into the formula.

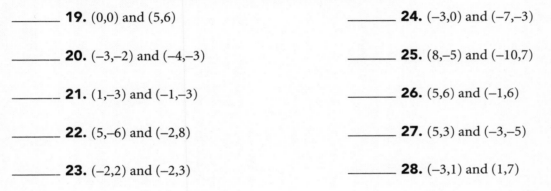

$$\begin{array}{cc} x_1 y_1 & x_2 y_2 \\ P(-6,5) & Q(-2,-1) \end{array}$$

$$\text{Slope} = \frac{y_2 - y_1}{x_2 - x_1} = \frac{-1 - 5}{-2 - (-6)} = \frac{-6}{-2 + 6} = \frac{-6}{4} = \frac{-3}{2}$$

Practice

Use a formula to find the slope of the line through each pair of points.

_____ **19.** $(0,0)$ and $(5,6)$

_____ **20.** $(-3,-2)$ and $(-4,-3)$

_____ **21.** $(1,-3)$ and $(-1,-3)$

_____ **22.** $(5,-6)$ and $(-2,8)$

_____ **23.** $(-2,2)$ and $(-2,3)$

_____ **24.** $(-3,0)$ and $(-7,-3)$

_____ **25.** $(8,-5)$ and $(-10,7)$

_____ **26.** $(5,6)$ and $(-1,6)$

_____ **27.** $(5,3)$ and $(-3,-5)$

_____ **28.** $(-3,1)$ and $(1,7)$

For questions 29–30, graph the points of the triangle and then calculate the slope of each side.

29. $\triangle STA$: $S(2,3)$; $T(4,-1)$; $A(2,-4)$

Slope of $\overline{AT}$: _____

Slope of $\overline{SA}$: _____

Slope of $\overline{ST}$: _____

30. $\triangle ABC$: $A(-4,-5)$; $B(-4,-1)$; $C(0,-5)$

Slope of $\overline{AC}$: _____

Slope of $\overline{AB}$: _____

Slope of $\overline{BC}$: _____

Skill Building until Next Time

Locate a set of steps near your home or school. Measure the rise and run of the set of steps. Use these measurements to find the slope of the steps.

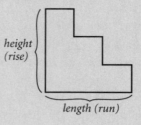

height (rise)

length (run)

19 ▶ THE EQUATION OF A LINE

LESSON SUMMARY

In this lesson, you will learn how to identify linear equations. You will also learn how to use the equation of a line to find points on the line. You will also determine if an ordered pair is on a line using the equation of the line.

People often make comparisons among numbers. For example, a nurse looks at a patient's temperature over a period of time. A salesperson will compare commissions from sports cars and family vans. These types of comparisons can often be graphed in a straight line to make predictions about future events. This straight line may be created by using a linear equation.

What Is a Linear Equation?

The standard form for a linear equation is $Ax + By = C$, where A, B, and C are constants; A and B cannot both be equal to zero. Linear equations will not have exponents on the variables x and y. The product or quotient of variables is not found in linear equations. Take a look at the following examples of linear and nonlinear equations.

Linear equations

$2x + 3y = 7$

$x = -3$

$y = \frac{2}{3}x + 4$

$y = 7$

Nonlinear equations

$2x^2 + 3y = 7$

$\frac{-3}{y} = x$

$y^2 = \frac{2}{3}x + 4$

$xy = 7$

Practice

Identify each equation as linear or nonlinear.

1. $x - 3y = 0$

2. $3y^2 = 2x$

3. $x^2 + y^2 = 49$

4. $x + y = 15$

5. $y = -12$

6. $\frac{1}{y} = 2$

7. $\frac{1}{3}x = 36$

8. $4x - 7y = 28$

9. $3x + 2y^2 = 12$

10. $x = 4y - 3$

Points on a Line

When you graph multiple points on a coordinate plane, you can easily see whether they could be connected to form a straight line. However, it is possible for you to determine whether a point is on a line or satisfies the equation of a line without using a coordinate plane. If the ordered pair can replace x and y, and the result is a true statement, then the ordered pair is a point on the line.

Examples: Determine if the ordered pairs satisfy the linear equation $2x - y = 4$.

(a) $(2,2)$ **Solution:** $2x - y = 4$

$$2(2) - 2 \overset{?}{=} 4$$
$$4 - 2 \overset{?}{=} 4$$
$$2 \neq 4$$

no; $(2,2)$ is not on the line $2x - y = 4$.

(b) $(0,-4)$ **Solution:** $2x - y = 4$

$$2(0) - (-4) \overset{?}{=} 4$$
$$0 + 4 \overset{?}{=} 4$$
$$4 = 4$$

yes; $(0,-4)$ is on the line $2x - y = 4$.

Practice

Determine if the ordered pairs satisfy the linear equation $2x + 5y = 10$.

11. $(0,-2)$

12. $(10,1)$

13. $(\frac{5}{2},3)$

14. $(-5,-4)$

15. $(15,4)$

16. $(3,-1)$

17. $(-\frac{5}{2},-3)$

18. $(5,0)$

Determine whether each ordered pair satisfies the linear equation $y = -2(2x - 3)$.

19. $(-3,-6)$

20. $(-1,-10)$

21. $(0,6)$

22. $(6,-18)$

23. $(2,2)$

24. $(3,-6)$

Graphing Linear Equations

You can graph any linear equation you want by choosing several x or y coordinates to substitute into the original equation, and then solve the equation to find the other variable. This is easier if you organize your work into a table. You could choose any values you want and get a solution that would be an ordered pair or point on the line, but for this example, let's use the x-values.

Example: Find three ordered pairs that satisfy the equation. Graph each equation.

$y = x + 4$

Solution:

x	$x + 4$	y	(x,y)
–2			
0			
2			

x	$x + 4$	y	(x,y)
–2	–2 + 4	2	(–2,2)
0	0 + 4	4	(0,4)
2	2 + 4	6	(2,6)

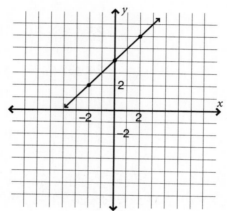

Practice

Find three ordered pairs that satisfy each equation. Then graph each equation.

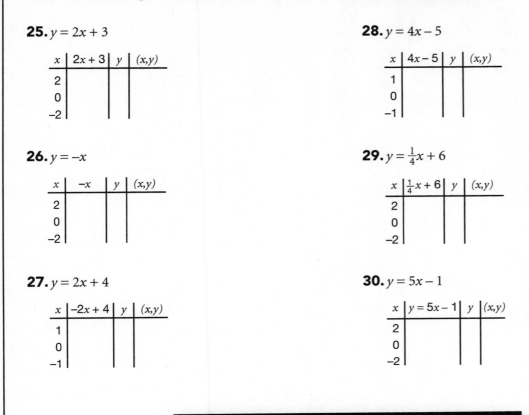

25. $y = 2x + 3$

x	2x + 3	y	(x,y)
2			
0			
-2			

26. $y = -x$

x	-x	y	(x,y)
2			
0			
-2			

27. $y = 2x + 4$

x	-2x + 4	y	(x,y)
1			
0			
-1			

28. $y = 4x - 5$

x	4x - 5	y	(x,y)
1			
0			
-1			

29. $y = \frac{1}{4}x + 6$

x	$\frac{1}{4}x + 6$	y	(x,y)
2			
0			
-2			

30. $y = 5x - 1$

x	y = 5x - 1	y	(x,y)
2			
0			
-2			

Skill Building until Next Time

The distance that a lightning flash is from you and the time in seconds that it takes for you to hear the thunder clap can be written in the following linear equation: $d = \frac{t}{5}$. Find at least three ordered pairs (t,d) that satisfy the formula, then graph the line.

20 ▶ TRIGONOMETRY BASICS

LESSON SUMMARY

In this lesson, you will learn how to write the trigonometric ratios sine, cosine, and tangent for a right triangle. You will learn how to use the trigonometry ratios to find unknown angle or side measurements.

Properties of similar right triangles are the basis for trigonometry. Measurements sometimes cannot be made by using a measuring tape. For example, the distance from a ship to an airplane can't be determined this way; however, it could be found by using trigonometric ratios. The word *trigonometry* comes from the Greek language; it means "triangle measurement."

Recall that the hypotenuse is the side of the triangle across from the right angle. The other two sides of the triangle are called legs. These two legs have special names in relation to a given angle: adjacent leg and opposite leg. The word *adjacent* means *beside*, so the adjacent leg is the leg beside the angle. The opposite leg is across from the angle. You can see in the following figure how the legs are named depending on which acute angle is selected.

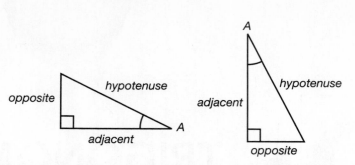

Sine, cosine, and tangent ratios for the acute angle A in a right triangle are as follows:

$\sin A = \frac{opposite\ leg}{hypotenuse}$

$\cos A = \frac{adjacent\ leg}{hypotenuse}$

$\tan A = \frac{opposite\ leg}{adjacent\ leg}$

Examples: Express each ratio as a decimal to the nearest thousandth.

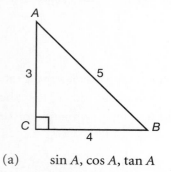

(a) $\sin A$, $\cos A$, $\tan A$

(b) $\sin B$, $\cos B$, $\tan B$

Solution:

(a) $\sin A = \frac{opp}{hyp} = \frac{4}{5} = 0.8$

$\cos A = \frac{adj}{hyp} = \frac{3}{5} = 0.6$

$\tan A = \frac{opp}{adj} = \frac{4}{3} \approx 1.3\overline{3}$

(b) $\sin B = \frac{opp}{hyp} = \frac{3}{5} = 0.6$

$\cos B = \frac{adj}{hyp} = \frac{4}{5} = 0.8$

$\tan B = \frac{opp}{adj} = \frac{3}{4} = 0.75$

Practice

Express each ratio as a decimal to the nearest thousandth. Use the following figure to answer practice problems 1–6.

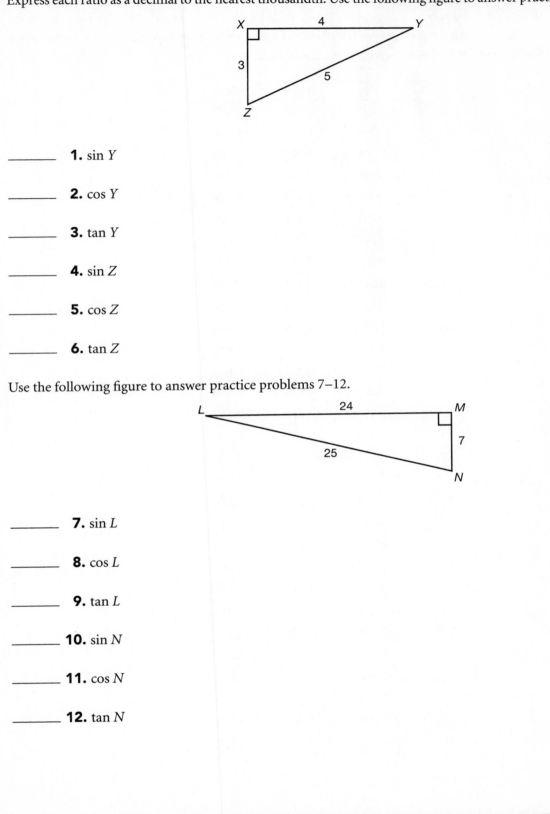

_____ **1.** sin Y

_____ **2.** cos Y

_____ **3.** tan Y

_____ **4.** sin Z

_____ **5.** cos Z

_____ **6.** tan Z

Use the following figure to answer practice problems 7–12.

_____ **7.** sin L

_____ **8.** cos L

_____ **9.** tan L

_____ **10.** sin N

_____ **11.** cos N

_____ **12.** tan N

Using a Trigonometric Table

Trigonometric ratios for all acute angles are commonly listed in tables. Scientific calculators also have functions for the trigonometric ratios. Consult your calculator handbook to make sure you have your calculator in the *degree* and not the *radian* setting. Part of a trigonometric table is given here.

ANGLE	SIN	COS	TAN
16°	0.276	0.961	0.287
17°	0.292	0.956	0.306
18°	0.309	0.951	0.325
19°	0.326	0.946	0.344
20°	0.342	0.940	0.364
21°	0.358	0.934	0.384
22°	0.375	0.927	0.404
23°	0.391	0.921	0.424
24°	0.407	0.914	0.445
25°	0.423	0.906	0.466
26°	0.438	0.899	0.488
27°	0.454	0.891	0.510
28°	0.470	0.883	0.532
29°	0.485	0.875	0.554
30°	0.500	0.866	0.577
31°	0.515	0.857	0.601
32°	0.530	0.848	0.625
33°	0.545	0.839	0.649
34°	0.559	0.829	0.675
35°	0.574	0.819	0.700
36°	0.588	0.809	0.727
37°	0.602	0.799	0.754
38°	0.616	0.788	0.781
39°	0.629	0.777	0.810
40°	0.643	0.766	0.839
41°	0.656	0.755	0.869
42°	0.669	0.743	0.900
43°	0.682	0.731	0.933
44°	0.695	0.719	0.966
45°	0.707	0.707	1.000

Example: Find each value.
              (a) cos 44°   (b) tan 42°

Solution:

              (a) cos 44° = 0.719   (b) tan 42° = 0.900

Example: Find $m\angle A$.
              (a) sin A = 0.656   (b) cos A = 0.731

Solution:

              (a) $m\angle A$ = 41°   (b) $m\angle A$ = 43°

Practice

Use the trigonometric table or a scientific calculator to find each value to the nearest thousandth or each angle measurement rounded to the nearest degree.

_____ **13.** sin 18°   _____ **19.** sin A = 0.485

_____ **14.** sin 32°   _____ **20.** cos A = 0.743

_____ **15.** cos 27°   _____ **21.** tan A = 0.384

_____ **16.** cos 20°   _____ **22.** cos A = 0.788

_____ **17.** tan 36°   _____ **23.** sin A = 0.8829

_____ **18.** tan 17°   _____ **24.** tan A = 0.306

Finding the Measure of an Acute Angle

The trigonometric ratio used to find the measure of an acute angle of a right triangle depends on which side lengths are known.

Example: Find $m\angle A$.

Solution: The sin A involves the two lengths known.

$$\sin A = \frac{opp}{hyp} = \frac{5}{16} \approx 0.313$$

Note that the ≈ symbol is used in the previous solution because the decimal 0.313 has been rounded to the nearest thousandths place.

Using the sin column of a trigonometric table, you'll find:

sin 18° = 0.309 } difference 0.004
sin A ≈ 0.313
sin 19° = 0.326 } difference 0.013
Since 0.313 is closer to 0.309, $m\angle A \approx 18°$ to the nearest degree.

A scientific calculator typically includes a button that reads sin⁻¹, which means *inverse sine*, and can be used to find an angle measure when the sine is known.

Example: Find $m\angle C$.

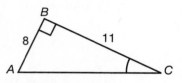

Solution: The tan C involves the two lengths known.

$$\tan C = \frac{opp}{adj} = \frac{8}{11} \approx 0.727$$

Using the tan column, you'll find:
tan 36 = 0.727
Therefore, $m\angle C = 36°$.

Again, the tan⁻¹ button on a scientific calculator can be used to find an angle measure when its tangent is known.

Practice
Find the measure of $\angle A$ to the nearest degree.

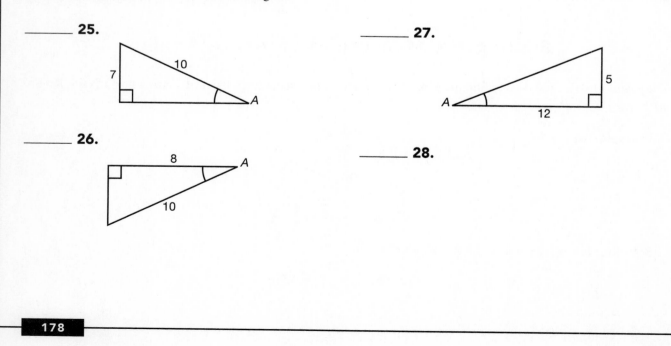

_____ **25.**

_____ **26.**

_____ **27.**

_____ **28.**

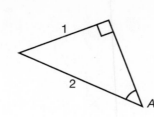

Finding the Measure of a Side

The trigonometric ratio used to find the length of a side of a right triangle depends on which side length and angle are known.

Example: Find the value of x to the nearest tenth.

(a)

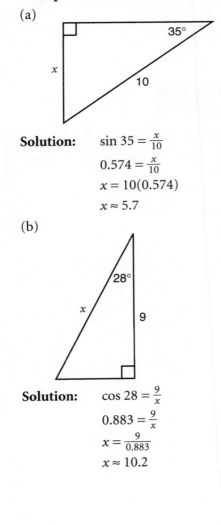

Solution: $\sin 35 = \frac{x}{10}$

$0.574 = \frac{x}{10}$

$x = 10(0.574)$

$x \approx 5.7$

(b)

Solution: $\cos 28 = \frac{9}{x}$

$0.883 = \frac{9}{x}$

$x = \frac{9}{0.883}$

$x \approx 10.2$

Practice

Find the value of *x* to the nearest tenth.

_____ **29.**

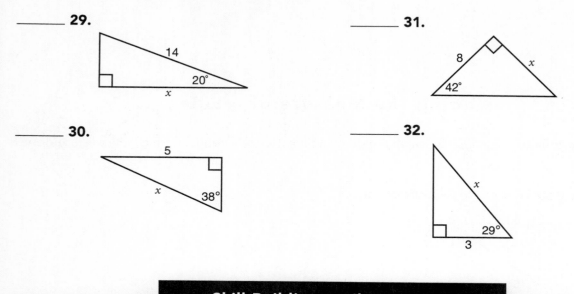

_____ **30.**

_____ **31.**

_____ **32.**

Skill Building until Next Time

Find a set of steps near your home or school. Measure the length of the hypotenuse and the adjacent leg. Then use cosine to find the angle of elevation of the steps. There are building codes that limit the angle of elevation of a set of steps to help prevent accidents.

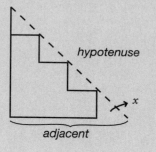

hypotenuse

adjacent

POSTTEST

Now that you have completed all the lessons, it is time to show off your new skills. Take the posttest in this chapter to see how much your geometry skills have improved. The posttest has 50 multiple-choice questions covering the topics you studied in this book. While the format of the posttest is similar to that of the pretest, the questions are different.

After you complete the posttest, check your answers with the answer key at the end of the book. Noted along with each answer is the lesson of this book that teaches you about the geometry skills needed for that question. If you still have weak areas, go back and work through the applicable lessons again.

1.	ⓐ	ⓑ	ⓒ	ⓓ
2.	ⓐ	ⓑ	ⓒ	ⓓ
3.	ⓐ	ⓑ	ⓒ	ⓓ
4.	ⓐ	ⓑ	ⓒ	ⓓ
5.	ⓐ	ⓑ	ⓒ	ⓓ
6.	ⓐ	ⓑ	ⓒ	ⓓ
7.	ⓐ	ⓑ	ⓒ	ⓓ
8.	ⓐ	ⓑ	ⓒ	ⓓ
9.	ⓐ	ⓑ	ⓒ	ⓓ
10.	ⓐ	ⓑ	ⓒ	ⓓ
11.	ⓐ	ⓑ	ⓒ	ⓓ
12.	ⓐ	ⓑ	ⓒ	ⓓ
13.	ⓐ	ⓑ	ⓒ	ⓓ
14.	ⓐ	ⓑ	ⓒ	ⓓ
15.	ⓐ	ⓑ	ⓒ	ⓓ
16.	ⓐ	ⓑ	ⓒ	ⓓ
17.	ⓐ	ⓑ	ⓒ	ⓓ

18.	ⓐ	ⓑ	ⓒ	ⓓ
19.	ⓐ	ⓑ	ⓒ	ⓓ
20.	ⓐ	ⓑ	ⓒ	ⓓ
21.	ⓐ	ⓑ	ⓒ	ⓓ
22.	ⓐ	ⓑ	ⓒ	ⓓ
23.	ⓐ	ⓑ	ⓒ	ⓓ
24.	ⓐ	ⓑ	ⓒ	ⓓ
25.	ⓐ	ⓑ	ⓒ	ⓓ
26.	ⓐ	ⓑ	ⓒ	ⓓ
27.	ⓐ	ⓑ	ⓒ	ⓓ
28.	ⓐ	ⓑ	ⓒ	ⓓ
29.	ⓐ	ⓑ	ⓒ	ⓓ
30.	ⓐ	ⓑ	ⓒ	ⓓ
31.	ⓐ	ⓑ	ⓒ	ⓓ
32.	ⓐ	ⓑ	ⓒ	ⓓ
33.	ⓐ	ⓑ	ⓒ	ⓓ
34.	ⓐ	ⓑ	ⓒ	ⓓ

35.	ⓐ	ⓑ	ⓒ	ⓓ
36.	ⓐ	ⓑ	ⓒ	ⓓ
37.	ⓐ	ⓑ	ⓒ	ⓓ
38.	ⓐ	ⓑ	ⓒ	ⓓ
39.	ⓐ	ⓑ	ⓒ	ⓓ
40.	ⓐ	ⓑ	ⓒ	ⓓ
41.	ⓐ	ⓑ	ⓒ	ⓓ
42.	ⓐ	ⓑ	ⓒ	ⓓ
43.	ⓐ	ⓑ	ⓒ	ⓓ
44.	ⓐ	ⓑ	ⓒ	ⓓ
45.	ⓐ	ⓑ	ⓒ	ⓓ
46.	ⓐ	ⓑ	ⓒ	ⓓ
47.	ⓐ	ⓑ	ⓒ	ⓓ
48.	ⓐ	ⓑ	ⓒ	ⓓ
49.	ⓐ	ⓑ	ⓒ	ⓓ
50.	ⓐ	ⓑ	ⓒ	ⓓ

Posttest

1. Which of the following is not a correct name for the line?

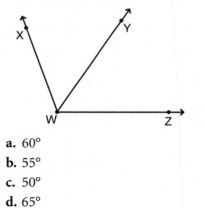

   a. $\overleftrightarrow{XZ}$
   b. $\overrightarrow{ZX}$
   c. $\overleftrightarrow{XYZ}$
   d. $\overrightarrow{YZ}$

2. What is the range of degrees for an acute angle?
   a. greater than 0, less than 90
   b. greater than 90, less than 180
   c. greater than 0, less than 180
   d. greater than 90, less than 360

3. In the figure, $\angle XWZ = 130°$ and $\angle YWZ = 70°$. What is the measure of $\angle XWY$?

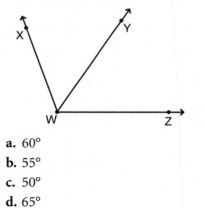

   a. 60°
   b. 55°
   c. 50°
   d. 65°

4. What is the measure of $\angle MOL$?

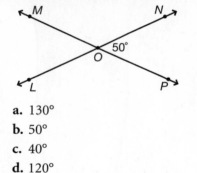

   a. 130°
   b. 50°
   c. 40°
   d. 120°

5. Classify ΔMNO with $\angle M = 25°$, $\angle N = 65°$, and $\angle O = 90°$.
   a. acute
   b. right
   c. obtuse
   d. equiangular

6. Which postulate would you use to prove $\Delta MNO \cong \Delta RST$?

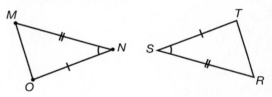

   a. AAA
   b. SAS
   c. ASA
   d. SSS

7. Three sides of a triangle are 6, 8, and 10; what type of triangle is it?
   a. acute
   b. right
   c. obtuse
   d. straight

8. Which of the following is not an acceptable name for the pentagon?

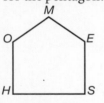

a. *MESHO*

b. *SEMOH*

c. *HEMOS*

d. *HOMES*

9. Which of the following is a property of rectangles?

a. The diagonals are congruent.

b. Opposite sides are congruent.

c. Opposite sides are parallel.

d. all of the above

10. Solve for x: $\frac{11}{25} = \frac{x}{100}$.

a. 44

b. 4

c. 11

d. 25

11. Find the perimeter of a rectangle with base 5 cm and height 20 cm.

a. 100 cm

b. 25 cm

c. 50 cm

d. 55 cm

12. Find the area of a triangle with a base that measures 35 feet and height 10 feet.

a. 350 ft.^{2}

b. 70 ft.^{2}

c. 3.5 ft.^{2}

d. 175 ft.^{2}

13. Find the surface area of a cube that measures 5 inches on an edge. Use $S.A. = 6e^2$.

a. 300 in.^{2}

b. 900 in.^{2}

c. 150 in.^{2}

d. 30 in.^{2}

14. Find the volume of the pyramid.

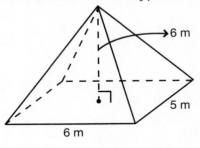

a. 60 m^3

b. 180 m^3

c. 30 m^3

d. 36 m^3

15. Find the volume of a cone with radius 9 cm and height 2 cm. Use 3.14 for π.

a. 508.68 cm^3

b. 3194.51 cm^3

c. 56.52 cm^3

d. 169.56 cm^3

16. In which quadrant will you graph the point $(-4,8)$?

a. I

b. II

c. III

d. IV

17. Find the slope of the line that passes through the points $(1,2)$ and $(3,3)$.

a. $\frac{4}{5}$

b. 2

c. $\frac{1}{2}$

d. $\frac{5}{4}$

18. Which ordered pair does not satisfy the equation $2x - y = 6$?

   a. (3,0)

   b. (0,−6)

   c. (1,−4)

   d. (2,6)

19. What is the sin B for this figure?

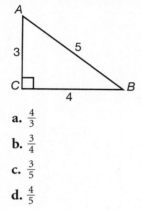

   a. $\frac{4}{3}$

   b. $\frac{3}{4}$

   c. $\frac{3}{5}$

   d. $\frac{4}{5}$

20. Which of the following sets of points are collinear?

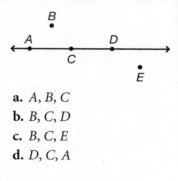

   a. A, B, C

   b. B, C, D

   c. B, C, E

   d. D, C, A

21. Which of the following is a pair of same-side interior angles?

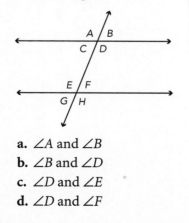

   a. $\angle A$ and $\angle B$

   b. $\angle B$ and $\angle D$

   c. $\angle D$ and $\angle E$

   d. $\angle D$ and $\angle F$

22. What is the correct classification for this angle?

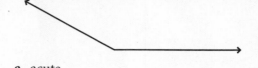

   a. acute

   b. right

   c. obtuse

   d. straight

23. What type of angle is shaped like a corner of a piece of paper?

   a. acute

   b. right

   c. obtuse

   d. straight

24. What is the measurement of $\angle AOC$?

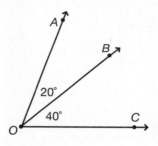

   a. 20°

   b. 80°

   c. 60°

   d. 85°

25. In the figure $\angle OMN = 45°$, what is the measurement of $\angle LMO$?

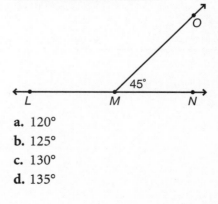

   a. 120°

   b. 125°

   c. 130°

   d. 135°

26. What is the measure of ∠A XB?

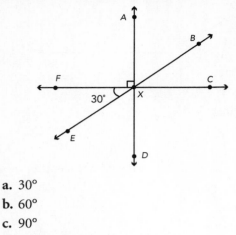

a. 30°

b. 60°

c. 90°

d. 180°

27. Classify the triangle in the figure by its angles.

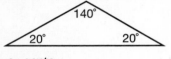

a. acute

b. right

c. obtuse

d. equiangular

28. What type of triangle is shown?

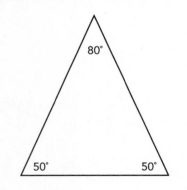

a. acute

b. right

c. obtuse

d. equiangular

29. Which postulate would you use to prove △ABD ≅ △CBD?

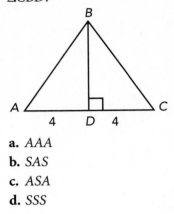

a. AAA

b. SAS

c. ASA

d. SSS

30. Find the missing length.

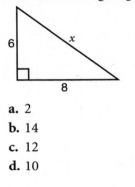

a. 2

b. 14

c. 12

d. 10

31. Three sides of a triangle are 7, 7, and 10; what type of triangle is it?

a. acute

b. right

c. obtuse

d. straight

32. Which figure is a concave polygon?

a.

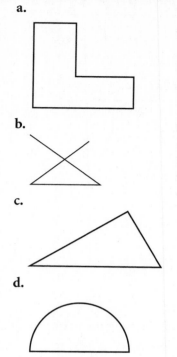

b.

c.

d.

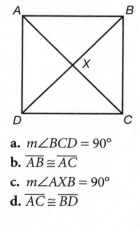

33. Which of the following is not a property of parallelograms?
a. Opposite sides of a parallelogram are congruent.
b. Opposite sides of a parallelogram are parallel.
c. Opposite angles of a parallelogram are congruent.
d. Diagonals of a parallelogram are congruent.

34. *ABCD* is a square. Which of the following is not true?

A ——— B

X

D ——— C

a. $m\angle BCD = 90°$
b. $\overline{AB} \cong \overline{AC}$
c. $m\angle AXB = 90°$
d. $\overline{AC} \cong \overline{BD}$

35. Name the means in the ratio: 3:4 = 6:8.
a. 3 and 4
b. 4 and 6
c. 3 and 8
d. 6 and 8

36. Find the perimeter of a regular pentagon that measures 3 ft. on each side.
a. 3 ft.
b. 9 ft.
c. 12 ft.
d. 15 ft.

37. Find the length of a side of an equilateral triangle whose perimeter is 42 m.
a. 126 m
b. 21 m
c. 14 m
d. 7 m

38. Find the area of the trapezoid in the figure.

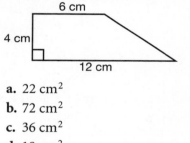

a. 22 cm²
b. 72 cm²
c. 36 cm²
d. 18 cm²

39. Find the surface area of a right rectangular prism with length 4 m, width 2 m, and height 8 m. Use the formula $S.A. = 2(lw + wh + lh)$.
a. 64 m²
b. 112 m²
c. 56 m²
d. 32 m²

40. How many faces does a cube have?
   a. 4
   b. 6
   c. 8
   d. 12

41. Find the area of the triangle.

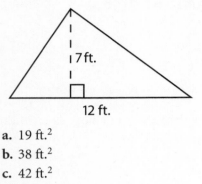

   a. 19 ft.^{2}
   b. 38 ft.^{2}
   c. 42 ft.^{2}
   d. 84 ft.^{2}

42. Find the volume of a cylinder with radius 5 cm and height 3 cm. Use 3.14 for π.
   a. 47.1 cm^3
   b. 78.5 cm^3
   c. 141.3 cm^3
   d. 235.5 cm^3

43. Find the surface area of a sphere with an 8-inch radius. Use 3.14 for π.
   a. 25.12 in.^{2}
   b. 631.01 in.^{2}
   c. 803.84 in.^{2}
   d. 200.96 in.^{2}

44. Find the distance between (1,2) and (3,2). Use $d = \sqrt{(x_2 - x_1)^2 + (y_2 - y_1)^2}$.
   a. 1
   b. 2
   c. 3
   d. 4

45. Find the slope of the line in the figure.

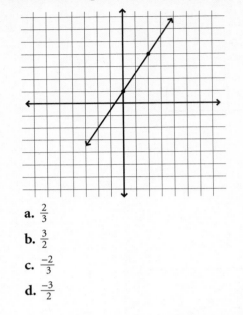

   a. $\frac{2}{3}$
   b. $\frac{3}{2}$
   c. $\frac{-2}{3}$
   d. $\frac{-3}{2}$

46. Find the volume of a rectangular prism with base area 50 m^2 and height 8 m.
   a. 100 m^3
   b. 200 m^3
   c. 400 m^3
   d. 800 m^3

47. If the x-value is 4, find the y-value in the equation $-x + y = 5$.
   a. 9
   b. 1
   c. 5
   d. −1

48. What is the trigonometric ratio for sine?
   a. $\frac{opposite\ leg}{hypotenuse}$
   b. $\frac{adjacent\ leg}{hypotenuse}$
   c. $\frac{opposite\ leg}{adjacent\ leg}$
   d. $\frac{adjacent\ leg}{opposite\ leg}$

49. In the following figure, what is the tan *A*?

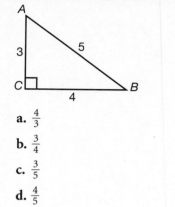

   a. $\frac{4}{3}$

   b. $\frac{3}{4}$

   c. $\frac{3}{5}$

   d. $\frac{4}{5}$

50. For the following figure, what is the correct sequence of the slopes of the various line segments from left to right?

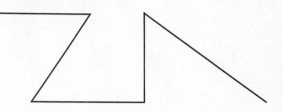

   a. underfined, negative, undefined, zero, positive

   b. zero, positive, zero, undefined, negative

   c. zero, positive, zero slope, undefined, positive

   d. undefined, positive, undefined, zero, negative

ANSWER KEY ▶

Pretest

If you miss any of the answers, you can find help for that kind of question in the lesson shown to the right of the answer.

1. d Lesson 1	**21. d** Lesson 9
2. a Lesson 1	**22. d** Lesson 9
3. c Lesson 1	**23. b** Lesson 9
4. c Lesson 2	**24. a** Lesson 10
5. d Lesson 2	**25. c** Lesson 10
6. d Lesson 3	**26. c** Lesson 11
7. b Lesson 3	**27. a** Lesson 11
8. a Lesson 3	**28. d** Lesson 11
9. d Lesson 4	**29. b** Lesson 12
10. c Lesson 4	**30. c** Lesson 12
11. b Lesson 5	**31. a** Lesson 13
12. b Lesson 5	**32. a** Lesson 13
13. d Lesson 5	**33. c** Lesson 13
14. a Lesson 6	**34. d** Lesson 14
15. b Lesson 6	**35. a** Lesson 14
16. c Lesson 6	**36. c** Lesson 15
17. b Lesson 7	**37. c** Lesson 15
18. d Lesson 7	**38. b** Lesson 15
19. c Lesson 8	**39. d** Lesson 16
20. c Lesson 8	**40. c** Lesson 16

41. d Lesson 17

42. b Lesson 17

43. a Lesson 17

44. a Lesson 18

45. d Lesson 18

46. d Lesson 19

47. d Lesson 19

48. d Lesson 19

49. c Lesson 20

50. b Lesson 20

Lesson 1

1. Yes; there are countless points on any line.

2. Points are distinguished from one another by the names assigned to them: *A*, *B*, *C*, and so on.

3. Lines, segments, rays, and planes are made up of a series of points.

4. $\overleftrightarrow{XY}, \overleftrightarrow{YZ}, \overleftrightarrow{XZ}, \overleftrightarrow{YX}, \overleftrightarrow{ZY}, \overleftrightarrow{ZX}$

5. An infinite number of points are on a line.

6. The notation for a line has two arrowheads because a line extends forever in both directions.

7. $\overrightarrow{SR}$; $\overrightarrow{ST}$

8. The endpoint is the beginning of a ray.

9. They are different because they have different endpoints and extend in different directions.

10. $\overline{LM}$; $\overline{MN}$; $\overline{NP}$; $\overline{LN}$; $\overline{MP}$; $\overline{LP}$

11. Line segments do not extend indefinitely. They have starting points and stopping points.

12. An infinite number of points are on a line segment.

13. A line has no endpoints; a ray has one endpoint.

14. A point has no size, has no dimension, indicates a definite location, and is named with an italicized capital letter.

15. A ray extends indefinitely in one direction, but a segment has two endpoints.

16. A line segment is part of a line, has endpoints, includes an infinite set of points, and is one-dimensional.

17. A plane has two dimensions; a line has one dimension.

18. Answers will vary but could include a sphere, cube, rectangular prism, or triangular prism.

19. Yes, a third point could be off the line.

20. Yes, coplanar points can be noncollinear because two points could be on one line with a third point that lies the same plane but not on the same line.

21. Yes, collinear points must be coplanar because if a line is in a plane, then all points on that line are in the same plane.

22. yes

23. no

24. yes

25. Yes; remember that any two points determine a line, even if it is not drawn.

26. yes

27. no

28. no

29. Yes; remember that any three noncollinear points determine a plane, even if it is not drawn.

30. true

31. false

32. true

33. False; sometimes they do, but not always.

34. ←•——•———•—→

35. Not possible; any three points are coplanar.

36. always

Lesson 2

1. Vertex: *Y*; sides: $\overrightarrow{YX}$; $\overrightarrow{YZ}$

2. Vertex: *J*; sides: $\overrightarrow{JA}$; $\overrightarrow{JT}$

3. ∠*KBT*, ∠*TBK*, ∠*B*, ∠*1*

4. two rays with a common endpoint

5. ∠*PON* (∠*NOP*) and ∠*POM* (∠*MOP*)

6. ∠*MON* (∠*NOM*)

7. $\overrightarrow{ON}$ and $\overrightarrow{OM}$

8. More than one angle uses the letter *O* as a vertex; therefore, no one would know which ∠*O* you meant.

9. No; they don't share the same endpoint.

10. No; they may not form a line.

11. ∠*KOL*; ∠*JOK*; ∠*NOM*

12. ∠*NOK*; ∠*MOJ*; ∠*MOL*

13. ∠*NOL*; ∠*MOK*

14. ∠*MOK*; ∠*KOM*

15. 63° (180° −27° −90°)

16. right

17. straight

18. acute

19. obtuse

20. straight angle

21. obtuse angle

22. acute

23. straight

24. obtuse

25. right

26. acute

27. obtuse

28. right

29. straight

30. The angle measures, from smallest to largest, are ∠*A*, ∠*C*, ∠*B*, and ∠*D*.

Lesson 3

1. Yes; line *d* cuts across lines *t* and *r*.

2. Yes; line *y* cuts across lines *t* and *r*.

3. No; line *t* intersects lines *d* and *y* at the same point, not two different points.

4. Yes; line *r* cuts across lines *d* and *y* at two different points.

5. true

6. False; the symbol means perpendicular.

7. False; transversals can be perpendicular, but they do not have to be perpendicular.

8. false

9. always

10. never

11. sometimes

12. never

13. never

14. never

15. corresponding angles; congruent

16. same-side interior angles; supplementary

17. corresponding angles; congruent

18. alternate interior angles; congruent

19. 3, 6, 8, 9, 11, 14, and 16

20. 2, 4, 5, 7, 10, 12, 13, and 15

21. 80°

22. 109°

23. 110°

24. 55°

25. They are both pairs of congruent angles when formed by parallel lines and a transversal.

26. Alternate interior angles are both "inside" the parallel lines. Corresponding angles are a pair of angles with one angle "inside" the parallel lines and one "outside" the parallel lines.

27. They are both pairs of interior angles.

28. Same-side interior angles are supplementary. Alternate interior angles are congruent.

29. Both pairs of angles are on the same side of the transversal.

30. Same-side interior angles are supplementary. Corresponding angles are congruent. Also, both same-side interior angles are interior while corresponding angles have one angle "inside" and one "outside" the parallel lines.

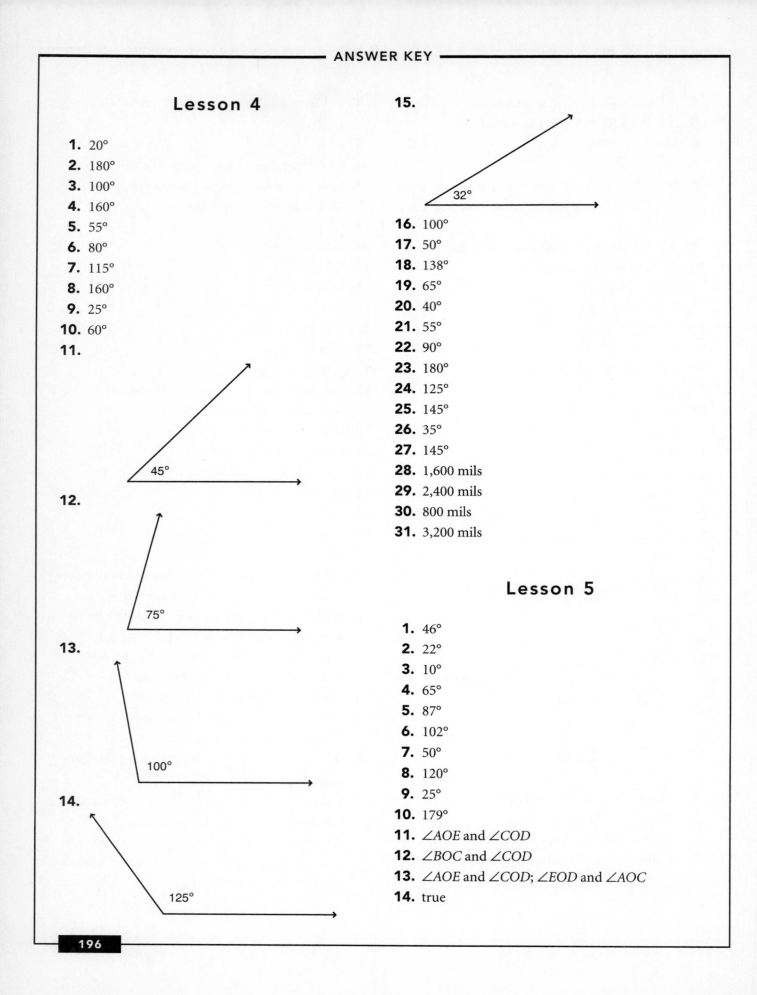

Lesson 4

1. 20°
2. 180°
3. 100°
4. 160°
5. 55°
6. 80°
7. 115°
8. 160°
9. 25°
10. 60°
11.

12.

13.

14.

15.

16. 100°
17. 50°
18. 138°
19. 65°
20. 40°
21. 55°
22. 90°
23. 180°
24. 125°
25. 145°
26. 35°
27. 145°
28. 1,600 mils
29. 2,400 mils
30. 800 mils
31. 3,200 mils

Lesson 5

1. 46°
2. 22°
3. 10°
4. 65°
5. 87°
6. 102°
7. 50°
8. 120°
9. 25°
10. 179°
11. ∠AOE and ∠COD
12. ∠BOC and ∠COD
13. ∠AOE and ∠COD; ∠EOD and ∠AOC
14. true

15. false

16. false

17. true

18. true

19. true

20. false

21. true

22. true

23. false

24. false

25. true

26. ∠2 and ∠5; ∠3 and ∠5; ∠1 and ∠4

27. ∠*XAW* and ∠*YAU*

28. ∠*XAZ* and ∠*ZAX*

29. ∠*XAY* and ∠*WAU*; ∠*XAW* and ∠*YAU*

30. 40°

Lesson 6

1. equilateral

2. isosceles

3. scalene

4. scalene

5. scalene

6. isosceles

7. scalene

8. equilateral

9. legs: $\overline{XY}$ and $\overline{YZ}$

vertex angle: ∠*Y*

base angles: ∠*X* and ∠*Z*

base: $\overline{XZ}$

10. legs: $\overline{RS}$ and $\overline{ST}$

vertex angle: ∠*S*

base angles: ∠*R* and ∠*T*

base: $\overline{RT}$

11. legs: $\overline{DE}$ and $\overline{EF}$

vertex angle: ∠*E*

base angles: ∠*D* and ∠*F*

base: $\overline{DF}$

12. acute

13. equiangular and acute

14. right

15. obtuse

16. obtuse

17. acute

18. right

19. false

20. true

21. true

22. false

23. false

24. true

25. true

26. false

27. true

28. base

29. acute, equiangular, right, obtuse, isosceles, equilateral, and scalene

30. a congruent angle

Lesson 7

1. ∠*R*

2. ∠*J*

3. ∠*K*

4. $\overline{PQ}$

5. $\overline{MK}$

6. $\overline{JM}$

7. Δ*FJH*

8. Δ*HGF*

9. Δ*FHG*

10. Δ*GFH*

11. $\overline{RS}$

12. $\overline{VS}$

13. $\overline{TR}$

14. yes

15. no

16. vertical angles

17. yes

18. $\overline{AC}$

19. $\overline{CD}$

20. no (should be $\triangle ACB \cong \triangle ECD$)

21. $\angle CDA$

22. $\overline{DB}$

23. no

24. yes

25. yes

26. SAS

27. ASA

28. ASA

29. SSS

30. SAS

Lesson 8

1. $\overline{AB}$ and $\overline{BC}$

2. $\overline{AC}$

3. $\overline{PR}$ and $\overline{RQ}$

4. $\overline{PQ}$

5. $\overline{HL}$ and $\overline{LM}$

6. $\overline{HM}$

7.

Number	Square	Number	Square
1	1	9	81
2	4	10	100
3	9	11	121
4	16	12	144
5	25	13	169
6	36	14	196
7	49	15	225
8	64	16	256

8. 25

9. 9

10. 8

11. 4

12. 178

13. yes

14. no

15. yes

16. no

17. yes

18. yes

19. right

20. acute

21. obtuse

22. obtuse

23. obtuse

24. acute

25. obtuse

26. $\overline{EI}$ = 40 cm, right triangle

27. $\overline{HK}$ = 45 ft., right triangle

28. When the ladder is placed 5 feet from the building, the ladder extends a little over 17 feet up the building. When the ladder is placed 3 feet from the building, it reaches about $17\frac{3}{4}$ feet up the building. It would be impractical to place the ladder that close or even closer because it would not be stable. You would not go very far up the ladder before you would be falling back down!

29. The ladder would need to be 15 feet from the base of the wall. ($15^2 + 20^2 = 25^2$)

30. The ladder would need to be 10 feet tall. ($6^2 + 8^2 = 10^2$)

Lesson 9

1. not a polygon

2. convex polygon

3. concave polygon

4. convex polygon

5. not a polygon

6. convex polygon

7. concave polygon

8. not a polygon

9. not a polygon

10. quadrilateral, yes

11. triangle, yes

12. pentagon, no

13. heptagon, no

14. hexagon, no

15. octagon, yes

16. yes

17. no

18. yes

19. quadrilateral, convex

20. yes

21. no

22. no

23. hexagon, concave

24. 540°

25. 900°

26.

Number of sides	6	10	14	16
Interior ∠ sum	720	1,440	2,160	2,520
Exterior ∠ sum	360	360	360	360

27. Home plate is a pentagon, so use the formula with $n = 5$.

$S = 180(n - 2)$

$S = 180(5 - 2)$

$S = 180(3)$

$S = 540$

3(right angles) + 2(obtuse angles) = 540

3(90) + 2(obtuse angles) = 540

270 + 2(obtuse angles) = 540

Subtract the 3 right angles to see how much of the original 540 is left.

2(obtuse angles) = 540 − 270 = 270

obtuse angle = 135 (divide by 2 because the two angles share the 270 evenly)

Therefore, each obtuse angle equals 135°. You can check to see that your answer is correct:

90 + 90 + 90 + 135 + 135 = 540.

28. $x = \frac{180(10 - 2)}{8} = 135°$

29. $x = \frac{180(10 - 2)}{10} = 144°$

30. $x = \frac{360}{6} = 60°$ (The sum of the exterior angles of a convex polygon is 360°.)

Lesson 10

1. true

2. false

3. true

4. false

5. true

6. true

7. false

8. false

9. false

10. false

11. 8

12. 12

13. 65°

14. 115°

15. 65°

16. 14

17. 5

18. 55°

19. 125°

20. 125°

21. 30

22. 15

23. 25°

24. 65°

25. 8

26. 28

27. 58°

28. 64°

29. 90°

30. 58°

31. 116°

32. 32°

33. Parallelograms and trapezoids are both quadrilaterals, but a trapezoid can never be a parallelogram or vice versa. Rhombuses and rectangles are parallelograms, with squares being both rhombuses and rectangles. Isosceles trapezoids are a special type of trapezoid.

Lesson 11

1. $\frac{10}{5} = \frac{2}{1}$
2. *KB:BT*
3. $\frac{5}{10} = \frac{1}{2}$
4. *BT:KB*
5. $\frac{5}{15} = \frac{1}{3}$
6. $\frac{BT}{KT}$
7. yes
8. 7 and 4
9. no
10. 12 and 15
11. no
12. 3 and 3
13. $x = 4$
14. $y = 44$
15. $z = 28$
16. $a = 8$
17. $b = 14$
18. $x = 1$
19. yes
20. no
21. no
22. yes
23. no
24. SAS
25. AA, SAS, or SSS
26. AA
27. AA
28. none
29. none
30. SAS
31. SSS

Lesson 12

1. 21
2. 36
3. 24
4. 33
5. 30
6. 22 in.
7. 21
8. 18 cm
9. 36 ft.
10. 26 cm
11. 80 cm
12. 30 ft.
13. 36 in.
14. 80 in.
15. 63 ft.
16. $w = 1$ m; $l = 5$ m
17. $w = 5$ cm; $l = 10$ cm
18. $s = 7$ in.
19. $s = 8$ yd.
20. $s = 20$ cm
21. $s = 10$ cm
22. $s = 8$ cm
23. $s = 23$ cm
24. 11 cm
25. 36 in.
26. 80 ft.
27. 68 ft.
28. 16 packages
29. 54 books (18 on each shelf)
30. 15 ft.

Lesson 13

1. 24 cm^2
2. 16 m^2
3. 196 in.^{2}
4. 21 ft.^{2}
5. $b = 8$ in.; $h = 7$ in.
6. $b = 15$ m; $h = 5$ m
7. $h = 6.25$ yd.
8. $b = 15$ cm
9. $b = 2$ cm, $h = 30$cm

10. $A = 24$ cm^2; $b = 8$ cm; $h = 3$ cm

11. $A = 41.5$ mm^2; $b = 8.3$ mm; $h = 5$ mm

12. $b = 6$ cm; $h = 10$ cm

13. $b = 7$ ft.; $h = 4$ ft.

14. $A = 21$ cm^2

15. $A = 30$ in.^{2}

16. $A = 35$ ft.^{2}

17. $A = 169$ in.^{2}

18. $A = 32$ cm^2

19. $h = 14$ cm

20. $h = 8$ ft.

21. $b = 17$ in.

22. 40 ft.

23. 8 in.

24. $A = 132$ m^2

25. $b_1 = 5$ cm

26. $A = 104$ in.^{2}

27. $h = 6$ in.

28. $A = 45$ in.^{2}

29. $b_2 = 9$ m

30. $h = 48$ in.

Lesson 14

1. 5 in.

2. 4 in.

3. 9 in.

4. 5 in.

5. 4 in.

6. 9 in.

7. 12 edges

8. 6 faces

9. 8 vertices

10. 4 vertices

11. 4 vertices

12. 3 edges

13. 22 in.^{2}

14. 78 ft.^{2}

15. 158 m^2

16. 200 cm^2

17. 174 in.^{2}

18. 220 ft.^{2}

19. $w = 7$ ft.

20. $h = 16$ m

21. 96 ft.^{2}

22. 216 cm^2

23. 24 in.^{2}

24. 294 m^2

25. 201.84 cm^2

26. $6e^2 = 486$; $e^2 = 81$; $e = 9$ ft.

27. 11 in.

28. 14 ft.

29. 912 in.^{2}

30. green

Lesson 15

1. 432 in.^{3}

2. 32 m^3

3. 343 cm^3

4. 240 ft.^{3}

5. 108 m^3

6. 100 ft.^{3}

7. 120 cm^3

8. 150 m^3

9. 36 in.^{3}

10. 1,120 cm^3

11. 576 ft.^{3}

12. 680 m^3

13. 9 cm

14. 7 in.

15. 1,000 m^3

16. 27 ft.^{3}

17. 64 in.^{3}

18. 729 cm^3

19. $e = 15$ ft.

20. 192 cm^3

21. 180 m^3

22. 400 in.^{3}

23. 60 ft.^{3}

24. 270 m^3

25. 85 cm^3

26. 128 in.^{3}

27. 416 ft.^{3}

28. 1 cm^3

29. 6 m

30. 9 ft.

Lesson 16

1. 47.1 ft.

2. 69.08 in.

3. 21.98 m

4. 157 m

5. $c = 25.12$ cm

6. 78.5 ft.^{2}

7. 200.96 m^2

8. 530.66 cm^2

9. 452.16 in.^{2}

10. 12.56 cm^2

11. 81.64 cm^2

12. 747.32 in.^{2}

13. 828.96 ft.^{2}

14. 678.24 cm^2

15. 1,271.7 cm^3

16. 339.12 m^3

17. 339.12 ft.^{3}

18. 4,559.28 ft.^{2}

19. 100.48 m^3

20. 340.17 in.^{3}

21. 452.16 cm^3

22. 256.43 ft.^{3}

23. 615.44 m^2

24. 803.84 in.^{2}

25. 3,215.36 cm^2

26. 5,024 ft.^{2}

27. 523.3 m^3

28. 1,436.03 in.^{3}

29. 7,234.56 m^3

30. 381.51 ft.^{3}

Lesson 17

1.–12.

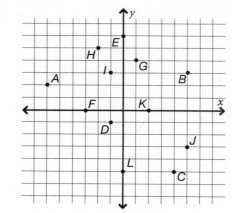

13. II

14. I

15. IV

16. III

17. y-axis

18. x-axis

19. I

20. The origin point (0,0) rests on both areas and is not in any specific quadrant.

21. III

22. IV

23. x-axis

24. y-axis

25. $A(-4,2)$

26. $B(1,-5)$

27. $C(4,0)$

28. $D(-6,-4)$

29. $E(-2,6)$

30. $F(0,-2)$

31. $G(6,-1)$

32. $H(3,4)$

33. 4

34. 10

35. 5

36. 13

Lesson 18

1. $\frac{1}{3}$

2. $\frac{2}{-1} = -2$

3. $\frac{4}{-1} = -4$

4. $\frac{3}{2}$

5. $\frac{2}{8} = \frac{1}{4}$

6. $\frac{2}{-5}$

7. $\frac{-2}{5}$

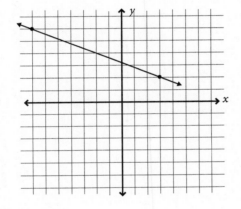

8. $\frac{-5}{2}$

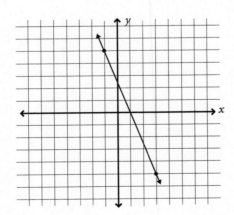

9. positive

10. negative

11. zero

12. positive

13. undefined (no slope)

14. negative

15. negative, zero, positive, undefined

16. zero, positive, zero, negative, zero

17. negative, positive, negative, zero, undefined

18. undefined, zero, undefined, zero, positive

19. $\frac{6}{5}$

20. 1

21. 0

22. -2

23. undefined (no slope)

24. $\frac{3}{4}$

25. $-\frac{2}{3}$

26. 0

27. 1

28. $\frac{6}{4}$

29. slope $AT = \frac{3}{2}$; slope $SA =$ undefined; slope $ST = -2$

30. slope $AC = 0$; slope $AB =$ undefined; slope $BC = -1$

Lesson 19

1. linear

2. nonlinear

3. nonlinear

4. linear

5. linear

6. linear

7. linear

8. linear

9. nonlinear

10. linear

11. yes

12. no

13. no

14. yes
15. yes
16. no
17. yes
18. yes
19. no
20. no
21. yes
22. yes
23. no
24. yes

25.

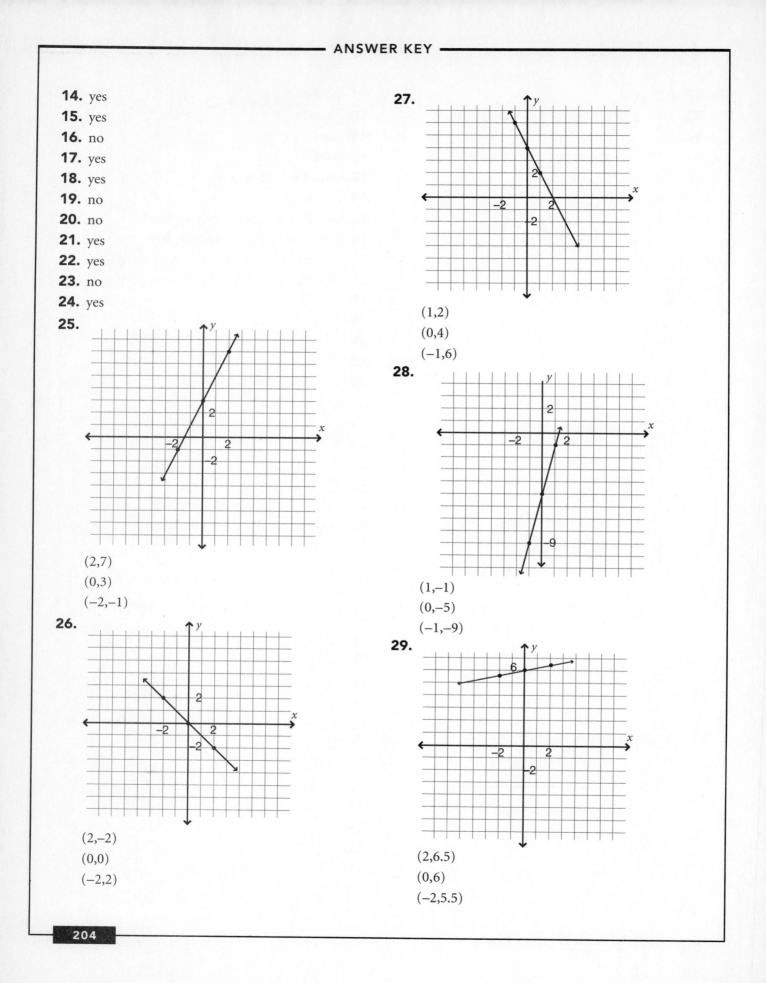

(2,7)
(0,3)
(−2,−1)

26.

(2,−2)
(0,0)
(−2,2)

27.

(1,2)
(0,4)
(−1,6)

28.

(1,−1)
(0,−5)
(−1,−9)

29.

(2,6.5)
(0,6)
(−2,5.5)

30.

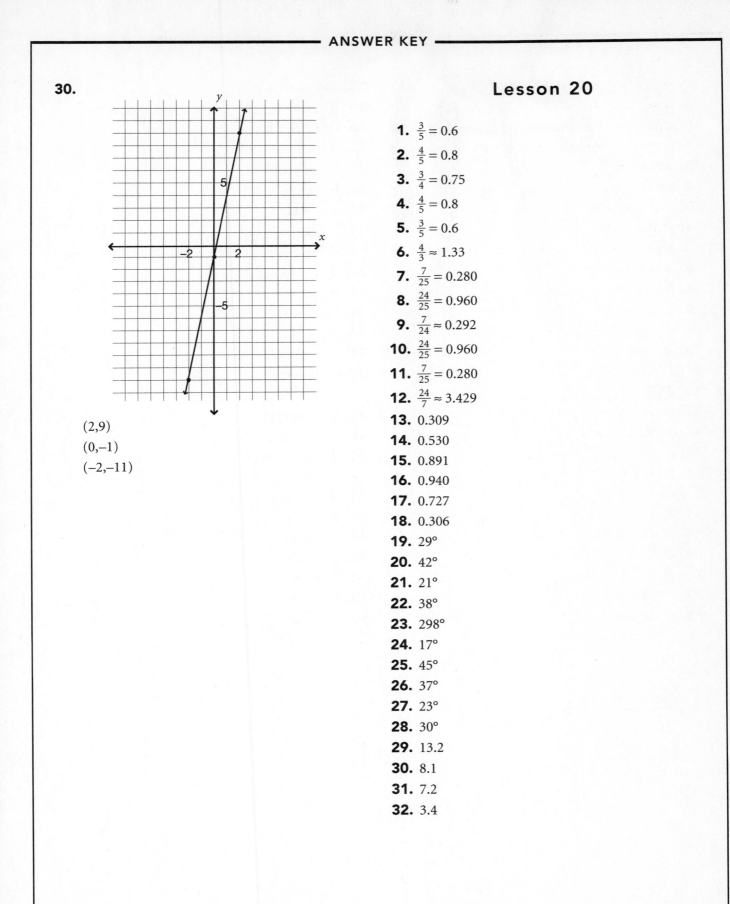

(2,9)
(0,−1)
(−2,−11)

1. $\frac{3}{5} = 0.6$

2. $\frac{4}{5} = 0.8$

3. $\frac{3}{4} = 0.75$

4. $\frac{4}{5} = 0.8$

5. $\frac{3}{5} = 0.6$

6. $\frac{4}{3} \approx 1.33$

7. $\frac{7}{25} = 0.280$

8. $\frac{24}{25} = 0.960$

9. $\frac{7}{24} \approx 0.292$

10. $\frac{24}{25} = 0.960$

11. $\frac{7}{25} = 0.280$

12. $\frac{24}{7} \approx 3.429$

13. 0.309

14. 0.530

15. 0.891

16. 0.940

17. 0.727

18. 0.306

19. 29°

20. 42°

21. 21°

22. 38°

23. 298°

24. 17°

25. 45°

26. 37°

27. 23°

28. 30°

29. 13.2

30. 8.1

31. 7.2

32. 3.4

Posttest

If you miss any of the answers, you can find help in the lesson shown to the right of the answer.

1. c   Lesson 1
2. a   Lesson 2
3. a   Lesson 4
4. b   Lesson 5
5. b   Lesson 6
6. b   Lesson 7
7. b   Lesson 8
8. c   Lesson 9
9. d   Lesson 10
10. a   Lesson 11
11. c   Lesson 12
12. d   Lesson 13
13. c   Lesson 14
14. a   Lesson 15
15. d   Lesson 16
16. b   Lesson 17
17. c   Lesson 18
18. d   Lesson 19
19. c   Lesson 20
20. d   Lesson 1
21. d   Lesson 3
22. c   Lesson 2
23. b   Lesson 2

24. c   Lesson 4
25. d   Lesson 4
26. b   Lesson 5
27. c   Lesson 6
28. a   Lesson 6
29. b   Lesson 7
30. d   Lesson 8
31. c   Lesson 8
32. a   Lesson 9
33. d   Lesson 10
34. b   Lesson 10
35. b   Lesson 11
36. d   Lesson 12
37. c   Lesson 12
38. c   Lesson 13
39. b   Lesson 14
40. b   Lesson 14
41. c   Lesson 13
42. d   Lesson 16
43. c   Lesson 16
44. b   Lesson 17
45. b   Lesson 18
46. c   Lesson 15
47. a   Lesson 19
48. a   Lesson 20
49. a   Lesson 20
50. b   Lesson 18

GLOSSARY

Acute angle: An angle that measures between 0° and 90°.

Acute triangle: A triangle with three acute angles.

Angle: A figure formed by two rays or two segments with a common endpoint.

Area: The amount of surface in a region. Area is expressed in square units.

Base angles of an isosceles triangle: The two angles that have the base as part of one side.

Base of a cone: Its circular face.

Base of an isosceles triangle: In an isosceles triangle with two congruent sides, the third side is called the base of the isosceles triangle.

Base of a trapezoid: Either of the parallel sides.

Circle: All points in a plane that are the same distance from some point called the center.

Circumference: The distance around a circle.

Collinear points: Points that lie on a line.

Complementary angles: Two angles whose measures total 90°.

Congruent (≅): Being the same or equal to.

Coordinates (*x*,*y*): An ordered pair of numbers used to name (locate) a point in a plane.

Coordinate plane: A plane that contains a horizontal number line (the *x*-axis) and a vertical number line (the *y*-axis). Every point in the coordinate plane can be named by a pair of numbers.

Coplanar points: Points that lie on a plane.

Corresponding parts: Any pair of sides or angles in congruent or similar polygons having the same relative position.

Cosine (cos): The ratio of the length of the leg adjacent to an acute angle of a right triangle to the length of the hypotenuse.

Cube: A rectangular prism in which all faces are congruent squares.

Diagonal of a polygon: A segment that joins two nonconsecutive vertices of the polygon.

Diameter of a circle: A segment that joins two points on the circle and passes through the center.

Edge: A segment formed by the intersection of two faces of a three-dimensional figure.

Equiangular triangle: A triangle with all angles equal.

Equilateral triangle: A triangle with all sides equal.

Extremes: The first and fourth terms of a proportion.

Face: A part of a plane forming a side of a three-dimensional figure.

Graph of an equation: The geometric figure that contains all the points whose coordinates make the equation a true statement.

Hexagon: A polygon with six sides.

Hypotenuse: The side opposite the right angle of a right triangle.

Intersecting lines: Lines that meet in one point.

Isosceles triangle: A triangle with two congruent sides.

Legs of a right triangle: Sides that determine the right angle.

Line ($\leftrightarrow$): A line is an undefined term for a set of points that extend indefinitely in two directions.

Linear equation: An equation whose graph is a line.

Means: The second and third terms of a proportion.

Measure of an angle: The number of degrees of an angle. Shown by the notation $m\angle ABC$.

Noncollinear points: A set of points through which a line cannot be drawn.

Noncoplanar points: A set of points through which a plane cannot be drawn.

Obtuse angle: An angle that measures between 90° and 180°.

Obtuse triangle: A triangle with one obtuse angle.

Octagon: A polygon with eight sides.

Opposite rays: Two rays that have a common endpoint and form a line.

Parallel lines ($\parallel$): Coplanar lines that do not intersect.

Parallelogram ($\square$): A quadrilateral with two pairs of opposite sides parallel. Rectangles, rhombuses, and squares are special types of parallelograms.

Pentagon: A polygon with five sides.

Perimeter: The distance around a two-dimensional figure.

Perpendicular ($\perp$): Lines, segments, or rays that intersect to form right angles.

Pi (π): The ratio of the circumference of a circle to its diameter. The most commonly used approximations for π are $\frac{22}{7}$ and 3.14.

Plane: An undefined term for a flat surface that extends indefinitely in all directions.

Point: An undefined term for a figure that indicates a definite location.

Polygon: A simple, closed, two-dimensional figure formed only by line segments that meet at points called vertices.

Polyhedron: A three-dimensional figure in which each surface is a polygon.

Postulate: A statement that is accepted without proof.

Proportion: A statement that two ratios are equal.

Protractor: An instrument used to measure angles.

Pythagorean theorem: For any right triangle, the square of the length of the hypotenuse is equal to the sum of the squares of the lengths of the legs.

Quadrant: One of the four regions, labeled I–IV, into which the coordinate axes divide a coordinate plane.

Quadrilateral: A polygon with four sides.

Radius of a circle: A segment joining the center of the circle to any point on the circle.

Radius of a sphere: A segment joining the center of the sphere to any point on the sphere.

Ratio: A comparison of two numbers by division. If $b \neq 0$, the ratio of a to b is denoted by $\frac{a}{b}$, $a{:}b$, or a to b.

Ray ($\rightarrow$): Part of a line with one endpoint. It extends indefinitely in one direction. A ray is named by its endpoint and any other point on the ray. The endpoint is named first.

Rectangle: A parallelogram with four right angles. A square is a special type of rectangle.

Regular polygon: A polygon in which all sides are congruent and all angles are congruent.

Rhombus: A parallelogram with four congruent sides. A square is a special type of rhombus.

Right angle: An angle that measures 90°.

Right triangle: A triangle that contains one right angle.

Rise: The vertical distance from a given point to a second given point.

Run: The horizontal distance from a given point to a second given point.

Scalene triangle: A triangle with no sides equal.

Segment: Part of a line with two endpoints. A segment is named by its endpoints.

Sides of an angle: The rays that meet to form an angle.

Sine (sin): The ratio of the length of the leg opposite an acute angle of a right triangle to the length of the hypotenuse.

Skew lines: Two lines that are not coplanar and do not intersect.

Slope of a line: The measure of the steepness of a line. It is also the ratio of rise to run. Also, if a line contains points (x_1, x_2) and (y_1, y_2), then the slope of the line is $\frac{y_2 - y_1}{x_2 - x_1}$, provided $x_1 \neq x_2$.

Sphere: A figure in space whose points are the same distance from a particular point. This point is called the center.

Square: A parallelogram with four right angles and four congruent sides. A square can also be defined as a parallelogram that is both a rectangle and a rhombus.

Square of a number (n^2): A number multiplied by itself.

Square root ($\sqrt{}$): The positive number which when multiplied by itself gives the original number as a product. The square root of 25 is 5 because $5 \times 5 = 25$.

Straight angle: An angle that measures 180°.

Supplementary angles: Two angles that together measure 180°.

Tangent (tan): The ratio of the length of the leg opposite an acute angle of a right triangle to the adjacent leg.

Theorem: A statement that can be proved.

Transversal: A line that intersects two or more coplanar lines at different points.

Trapezoid: A quadrilateral with exactly one pair of opposite sides parallel.

Triangle: A polygon with three sides.

Trigonometry: Mathematical principles based on the properties of similar right triangles.

Vertex: The intersection of two sides of a polygon.

Vertical angles: Two angles with sides forming two pairs of opposite rays. Vertical angles are congruent.

Volume: The amount of space in a solid. Volume is expressed in cubic units.

POSTULATES AND THEOREMS

Postulates

Lesson 1
- Two points determine exactly one line.
- Three noncollinear points determine exactly one plane.

Lesson 3
- If two parallel lines are cut by a transversal, then corresponding angles are congruent.

Lesson 4
- If point B is in the interior of $\angle AOC$, then $m\angle AOB + m\angle BOC = m\angle AOC$. (see page 41 for diagram)
- If $\angle AOC$ is a straight line, then $m\angle AOB + m\angle BOC = 180°$. (see page 41 for diagram)

Lesson 7
- If three sides of one triangle are congruent with three sides of another triangle, then the two triangles are congruent (SSS postulate).

- If two sides and the included angle of one triangle are congruent to the corresponding parts of another triangle, then the triangles are congruent (SAS postulate).
- If two angles and the included side of one triangle are congruent to corresponding parts of another triangle, the triangles are congruent (ASA postulate).

Lesson 11

- If two angles of one triangle are congruent to two angles of another triangle, then the triangles are similar (AA postulate).
- If the lengths of the corresponding sides of two triangles are proportional, then the triangles are similar (SSS postulate).
- If the lengths of two pairs of corresponding sides of two triangles are proportional and the corresponding included angles are congruent, then the triangles are similar (SAS postulate).

Theorems

Lesson 3

- If two parallel lines are cut by a transversal, then alternate interior angles are congruent.
- If two parallel lines are cut by a transversal, then same-side interior angles are supplementary.

Lesson 5

- Vertical angles are congruent.

Lesson 8

- Pythagorean theorem: In a right triangle, the sum of the squares of the lengths of the legs is equal to the square of the length of the hypotenuse ($a^2 + b^2 = c^2$).
- Converse of the Pythagorean theorem: If the square of the length of the longest side of a triangle is equal to the sum of the squares of the lengths of the two shorter sides, then the triangle is a right triangle.

- If the square of the length of the longest side is greater than the sum of the squares of the lengths of the other two shorter sides, then the triangle is obtuse ($c^2 > a^2 + b^2$). (see page 74 for diagram)
- If the square of the length of the longest side is less than the sum of the squares of the lengths of the two other sides, then the triangle is acute ($c^2 > a^2 + b^2$). (see page 74 for diagram)

Lesson 9

- The sum of interior angles of a triangle is 180°.
- If a convex polygon has n sides, then its angle sum is given by the formula $S = 180(n - 2)$.
- The sum of exterior angles of a convex polygon is always 360°.

Lesson 10

- Opposite sides of a parallelogram are congruent.
- Opposite angles of a parallelogram are congruent.
- Consecutive angles of a parallelogram are supplementary.
- Diagonals of a parallelogram bisect each other.
- Diagonals of a rectangle are congruent.
- The diagonals of a rhombus are perpendicular, and they bisect the angles of the rhombus.

Lesson 13

- The area (A) of a rectangle is the product of its base length (b) and its height (h): $A = bh$.
- The area of a parallelogram (A) is the product of its base length (b) and its height (h): $A = bh$.
- The area (A) of any triangle is half the product of its base length (b) and its height (h): $A = \frac{1}{2}bh$.
- The area of a trapezoid is half the product of the height and the sum of the base lengths ($b_1 + b_2$): $A = \frac{1}{2}bh(b_1 + b_2)$.

Lesson 14

- The surface area (*S.A.*) of a rectangular prism is twice the sum of the length (*l*) times the width (*w*), the width (*w*) times the height (*h*), and the length (*l*) times the height (*h*): $S.A. = 2(lw + wh + lh)$.
- The surface area of a cube is six times the edge (*e*) squared: $S.A. = 6e^2$.

Lesson 15

- To find the volume (*V*) of a rectangular prism, multiply the length (*l*) by the width (*w*) and by the height (*h*): $V = lwh$.
- To find the volume (*V*) of any prism, multiply the area of the base (*B*) by the height (*h*): $V = Bh$.
- The volume of a cube is determined by cubing the length of the edge: $V = e^3$.

Lesson 16

- The circumference of any circle is the product of its diameter and π: $C = \pi d$ or $C = 2\pi r$.
- The area (*A*) of a circle is the product of and the square of the radius (*r*): $A = \pi r^2$.
- The surface area (*S.A.*) of a cylinder is determined by finding the sum of the area of the bases and the product of the circumference times the height: $S.A. = 2\pi r^2 + 2\pi rh$.
- The volume (*V*) of a cylinder is the product of the area of the base (*B*) and the height (*h*): $V = Bh$ or $V = \pi r^2 h$.
- The surface area (*S.A.*) formula for a sphere is four times π times the radius squared: $S.A. = 4\pi r^2$.
- The volume (*V*) of a sphere is determined by the product of $\frac{4}{3}\pi$ times the radius cubed: $V = \frac{4}{3}\pi r^3$.

Lesson 17

- The distance *d* between any two points $A(x_1, y_1)$ and $B(x_2, y_2)$ is $d = \sqrt{(x_2 - x_1)^2 + (y_2 - y_1)^2}$.

B ▶ ADDITIONAL RESOURCES

Many resources are available to help you if you need additional practice with geometry. Your local high school is a valuable resource. A high school math teacher might assist you if you asked for help with a lesson, provide you with practice sets of problems on a lesson that proved difficult for you, or suggest a tutor for you. You could also check the classified ads in your local newspaper or the yellow pages to search for a tutor.

Colleges are also a valuable resource. They often have learning centers or tutor programs available. To find out what is available in your community, call your local college's math department or learning center.

If you would like to continue working geometry problems on your own, your local bookstore or library has books that can help you. You may also be able to borrow a textbook from your local high school. Check your local bookstore or library for availability.

ADDITIONAL ONLINE PRACTICE

Whether you need help building basic skills or preparing for an exam, visit the LearningExpress Practice Center! On this site, you can access additional practice materials. Using the code below, you'll be able to log in and take an additional geometry practice exam. This online practice exam will also provide you with:

- **Immediate Scoring**
- **Detailed answer explanations**
- **Personalized recommendations for further practice and study**

Log on to the LearningExpress Practice Center by using the URL: **www.learnatest.com/practice**

This is your Access Code: **7458**

Follow the steps online to redeem your access code. After you've used your access code to register with the site, you will be prompted to create a username and password. For easy reference, record them here:

Username: _____ **Password:** _____

With your username and password, you can log in and answer these practice questions as many times as you like. If you have any questions or problems, please contact LearningExpress customer service at 1-800-295-9556 ext. 2, or e-mail us at **customerservice@learningexpressllc.com**